NATURE'S THIRD CYCLE

NATURE'S THIRD CYCLE

A Story of Sunspots

Arnab Rai Choudhuri

OXFORD
UNIVERSITY PRESS

OXFORD
UNIVERSITY PRESS

Great Clarendon Street, Oxford, OX2 6DP,
United Kingdom

Oxford University Press is a department of the University of Oxford.
It furthers the University's objective of excellence in research, scholarship,
and education by publishing worldwide. Oxford is a registered trade mark of
Oxford University Press in the UK and in certain other countries

Published in the United States of America by Oxford University Press
198 Madison Avenue, New York, NY 10016, United States of America

British Library Cataloguing in Publication Data

Data available

Library of Congress Control Number: 2014942180

ISBN 978–0–19–967475–6

Printed in Great Britain by
Clays Ltd, St Ives plc

To Mahua

We meditate on the adorable glory of the radiant sun; may he inspire our intelligence.

—*Rigveda* 3.62.10 (translated by S. Radhakrishnan)

Foreword by Nigel Weiss

I welcome the opportunity of introducing this lively and unusual book. Sunspots are dark features on the surface of the Sun that have been observed through telescopes since the time of Galileo. Their incidence varies cyclically and the author focuses on this cycle, which is erratic and apparently chaotic, but nevertheless has an average period of 11 years that is well-defined. This is the 'third cycle' of his book. It contrasts with Nature's first two cycles—diurnal and annual—whose effects are familiar and whose periods scarcely change from one millennium to the next.

Sunspots are actually the sites of strong magnetic fields, generated by dynamo action in the solar interior and with systematic properties that reverse from one cycle to the next. Arnab Rai Choudhuri accepts the challenge of explaining this behaviour to non-experts, and he does so by intertwining this account with the tale of his own career. This scientific autobiography makes compulsive reading, and controversy is certainly not shirked. He began in India, in Kolkata, and obtained a graduate scholarship to the USA. In Chicago he found two giants in theoretical astrophysics. One was Subrahmanyan Chandrasekhar, who had found his way from India to Cambridge and thence to America; he was awarded a Nobel Prize in 1983 for predicting the formation of black holes—work done 50 years earlier. The other was Eugene Parker, who became Arnab's supervisor and inspired his career. At the time, Arnab was the only graduate student in this elite group. Parker had produced the first theoretical model of the solar dynamo; he also predicted the existence of the solar wind before it was ever detected; and he has published numerous fundamental papers on stellar and planetary magnetism. No wonder that he has remained a hero to Arnab ever since, as well as being a wise and generous adviser.

After he had graduated, Arnab made the brave decision to return to India and obtained a post at the Indian Institute of Science in Bangalore. At that time, research facilities in India were limited and funding was very short: for 12 years he had only a single opportunity of speaking at an international meeting, with support that was provided from abroad. But he did gain experience as a teacher and proved an outstanding expositor, as shown by the textbooks that he has written. As he gathered research students, he succeeded in publishing a series of key papers on the origin of the Sun's magnetic cycle. At the same time, India grew more prosperous and devoted funds to scientific research, so that there are now world-class facilities available to scientists there. Over time, he developed an international reputation and built up a flourishing research group, with talented students who have pursued their own careers in India and abroad.

The research output of this group has centred on modelling the dynamo process that generates cyclic activity in the Sun and other stars. In such models we can distinguish between two key components of the magnetic field: one is the 'toroidal' field, which winds azimuthally around the rotation axis, in or against the direction of the Sun's rotation; the other is the 'poloidal' field, which (in its simplest form) is that of a magnetic dipole aligned with the rotation axis and producing a field like that of an oscillatory bar magnet. It is easy to see how variations in the rotational velocity (such as are actually observed) can generate toroidal fields from poloidal fields—but the generation of a reversed poloidal field from the toroidal field is far more subtle. These processes, first described by Parker, are what Choudhuri calls the 'central dogma' of the solar cycle.

He and his colleagues have described the creation of isolated loops of magnetic flux deep inside the Sun, their interaction with rotation as they rise towards the surface, and their emergence to form sunspots. In their picture, these surface fields are progressively distorted to create a reversed poloidal field that is then transported towards the poles and downwards into deeper layers where a reversed toroidal field can be generated. Although this process remains controversial, Choudhuri achieved a triumph by accurately predicting that the current solar cycle would be much weaker than its immediate predecessors. His own account of the fierce arguments that followed at scientific meetings is scrupulously fair—but it does reveal the tensions that arose when others mistakenly insisted that the coming cycle would be much stronger than its predecessor.

Overall, Arnab Rai Choudhuri has succeeded in producing an interesting and instructive book. He writes in an engaging style and his own personal tale will carry readers into and through the scientific details, so that they emerge with a basic understanding of the problem of explaining the Sun's magnetic activity. I recommend it gladly and without any hesitation.

Nigel Weiss FRS
University of Cambridge
March 2014

Preface

Astrophysics is a branch of science which excites popular imagination and there has been a glorious tradition of working scientists themselves sometimes taking the lead in the popularization efforts. Pioneers like George Gamow and Fred Hoyle wrote virtually on all aspects of astrophysics as the subject was in their days. Of late, however, we see a tendency which I find disturbing. While popular science books keep being written on cosmology at an ever increasing rate, many other important areas of astrophysics have been grossly neglected by popular science writers in the past few years. There is no doubt that cosmology is one of the most thriving areas of astrophysics at the present time. But it simply does not make sense why serious efforts are not made to present many other important areas of astrophysics in interesting ways to the general reading public.

It is not that our sun does not get any media attention at all! Every now and then when we have solar disturbances affecting the earth, such events do get covered in the media. However, one usually finds a typical pattern in the coverage of these events in newspapers and magazines. Within the last few years, technological advances have enabled us to take spectacular images of solar events—both from the ground and from space. Usually newspaper or magazine coverage of a solar event would have spectacular photographs, with only some very superficial comments about the science behind these solar disturbances. The unwritten assumption would be that the underlying science is a kind of dry boring science which is best left to the experts: unlike the science behind cosmology or particle physics that can excite general readers. It is this assumption which I challenge in this book. The 11-year sunspot cycle is at the heart of all unusual solar events. I attempt to present the science behind this cycle as a part of the great intellectual tradition of physics. To the best of my knowledge, nobody has made such an attempt before me in a non-technical book. My firm belief that a fairly complete account of this science can be conveyed to non-experts (having some elementary knowledge of physics at the high school level) compelled me to write this book, against the advice of many well-wishers who felt that I was attempting something impossible. Only my readers can decide if I have succeeded and if they find this science interesting.

Having the good fortune of doing my PhD under arguably the greatest theoretical scientist in our field, I have intimately known many of the major players in this field and have taken part in some of the remarkable scientific advances over the past three decades. While telling the story of sunspots, I have sometimes taken the liberty of telling my story also (especially in Chapters 5, 7 and 9). I have found that most scientists are reticent in making their life stories public—perhaps due to the general perception that an average scientist's life is rather drab and colourless compared to an average artist's life. Through my story, I have tried to give an idea of how science is actually done in our field. I have tried to write about a scientist's hopes and fears, friendships, competitions, jealousies, and the utter joy of occasionally discovering a clue to understanding some deep mystery of nature. After spending several years in some top places in our field in the USA, I opted to work in a so-called 'third world country'. This personal experience has made me portray science as an intensely human activity in an uneven and heterogeneous planet.

Words fail me in thanking Nigel Weiss who enthusiastically wrote a Foreword for my book. Some friends who went through the entire manuscript very carefully and suggested numerous improvements are Subroto Mukerjee, D. P. Sengupta and L. Sharada. This would have been a much less satisfactory book without their valuable inputs. I am grateful to Ramesh Babu and Gopal Hazra for help with the figures. This project could not have been completed without the continuous support of my wife Mahua. Finally, I thank Sonke Adlung and Jessica White of Oxford University Press for their courage in publishing such a book by an author who had never written even a small piece of popular science in English before this book (my previous rare attempts being only in Bengali) and for their valuable guidance and encouragement.

Arnab Rai Choudhuri
April 2014

Contents

1

Explosions, Blackouts and Cycles

1.1 Explosions on the Sun

In the cold early-spring morning of 13 March 1989, six million people in Quebec in eastern Canada woke up to find that their homes were without electricity. The electricity could only be restored after about eight hours. In a developing country like India where I live and work, electricity companies are sometimes unable to meet the rapidly growing needs of power and occasionally stop supplying power to certain localities. But it is not very usual for citizens of Canada to go without electricity for eight hours. What could have caused this massive power blackout?

On 9 March 1989—four days before the power blackout—solar astronomers had noticed a gigantic explosion on the surface of the sun. Such explosions, which typically last for a few minutes, are called *solar flares*. A large solar flare can unleash energy which is close to a trillion (i.e. 10^{12}) times the energy of the atomic bombs which flattened Hiroshima and Nagasaki. Flares are undoubtedly amongst the most violent events taking place within the solar system. At the safe distance that we are, we can see a large flare on the sun's surface as a sudden brightening of a region which can typically be larger than the earth's size. Although it may at first seem improbable, the solar flare on the 9th of March appears to have been the cause of the power blackout in Quebec on the 13th of March!

The sun's rays gently warm our blue planet, allowing water to exist in the liquid state over much of the earth's surface and making life thrive in myriads of different forms. The earth would be a dead rocky place without this gentle warmth of the sun. But this gentle warmth comes from a steady flow of energy from the sun, which—as far as we can decipher—has remained fairly constant over the earth's long geological ages. Until about 150 years ago, nobody suspected that there can be any sudden occurrences in the sun that would affect the earth.

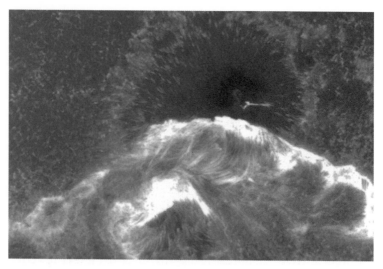

Figure 1.1 A solar flare seen above a large sunspot on 13 December 2006. This image was taken by the Solar Optical Telescope (SOT) on board the space mission Hinode. Credit: JAXA and NASA.

That there can be sudden explosions on the sun was discovered on 1 September 1859, by Richard Carrington, a gentleman astronomer who had set up a telescope at his home in Surrey, England. Carrington was then a young man of age 33 and his passion was to study sunspots—the dark patches on the surface of the sun. Some sunspots are so large that it would be possible to immerse the whole earth in one of them. We shall later have occasions to discuss some of the important discoveries Carrington made about sunspots. In the morning of 1 September 1859, Carrington was busy studying a sunspot with his telescope. It should be pointed out that an astronomer normally does not look at a sunspot directly through the telescope. That can seriously damage one's eyes. A typical procedure is to use the telescope to project an image of the sun onto a screen and to look for sunspots in that image. While studying the sunspot, Carrington suddenly found that 'two patches of intensely bright and white light broke out' in the middle of the sunspot. In his excitement to share this unusual spectacle with his friends, poor Carrington unfortunately missed the major part of this celestial display. He wrote:

> seeing the outburst to be very rapidly on the increase, and being some-
> what flurried by the surprise, I hastily ran to call some one to witness the

Figure 1.2 Richard Carrington's manor house in Surrey from where the first solar flare observation was made in 1859. The dome of the telescope can be seen. No photograph or portrait of Richard Carrington (1826–1875) is known to exist.

exhibition with me, and on returning within 60 seconds, was mortified to find that it was already much changed and enfeebled. Very shortly afterwards the last trace was gone, and although I maintained a strict watch for nearly an hour, no recurrence took place.[1]

When Carrington described this 'singular appearance seen in the sun' at the meeting of the Royal Astronomical Society two months later, he pointed out another curious thing. About 18 hours after he saw the flare, the magnetic observing station in Kew reported a sudden violent change in the earth's magnetic field, known as a *geomagnetic storm*. Carrington's observation that the solar flare and the geomagnetic storm occurred within a few hours of each other was recorded by the Royal Astronomical Society with the following cryptic remark:

While the contemporary occurrence may deserve noting, he would not have it supposed that he even leans towards hastily connecting them. 'One swallow does not make a summer.'

Well, that is the first indirect suggestion ever made by anybody that an explosion in the sun could cause peculiar things on the earth. Carrington's discovery raises several immediate questions. Firstly, sunspots are the darkest regions on the surface of the sun, whereas a flare is a sudden brightening. Since the flare occurred above the sunspot, one would think that the sunspot was in some way responsible for the flare. That means that darkest patches on the sun's surface can occasionally cause these remarkable sudden brightenings. This is literally a case of darkness creating light. Is this allowed by the laws of physics? Secondly, even if a flare is a very violent event, we should keep in mind that the sun keeps producing an enormous amount of energy steadily. A large flare, during its peak, may enhance the energy flux from the sun by only about one per cent. How could such a small temporary enhancement of the energy flux from the sun cause the blackout in Quebec? To get the full answers to these questions, you will have to read till Chapter 8. After all, when you read a detective novel, you usually have to read till the end to find out who the murderer is. Just as in a detective novel, you will find clues to answer these questions throughout this book.

The earth's magnetic field seems to provide a kind of shield protecting us from the effects of sudden solar disturbances, as we shall discuss in more detail in Chapter 8. This protective shield is weakest near the geomagnetic poles over which we do not have the canopy of overlying magnetic field lines. One geomagnetic pole is located in the Canadian Arctic. The region surrounding this pole is particularly vulnerable to solar disturbances and that is why the big blackout occurred in Quebec. Carrington's observation that there was a sudden change in the earth's magnetic field a few hours after he observed the 1859 flare gives us the physics clue to understand how electrical blackouts may be caused. One of the most fundamental discoveries in the history of electrical sciences is Michael Faraday's discovery in 1831 that a changing magnetic field can drive a current through a circuit placed in the region where the magnetic field is changing. A large power grid in regions near the geomagnetic pole can be thought of as a circuit. A sudden change in the earth's magnetic field can drive a huge current through this grid, thereby tripping it and causing a power blackout. We find many consequences of solar disturbances in the polar regions. The beautiful polar aurora, which is usually seen two or three days after a large flare, is one harmless consequence. However, the other consequences can be more troublesome. Some readers may know that the ionosphere reflects radio waves which are used for radio communication. That is why we can make radio contacts with distant places, which would otherwise not be possible because of the earth's curvature. A major flare can disturb the ionosphere, thereby causing disruptions in radio communication. Nowadays many airline companies have flights flying close to the north geomagnetic pole—especially for routes between Europe and North America. These routes become dangerous after a major solar flare. Certainly nobody would want the pilot to lose radio contact with the ground station or the electronics to malfunction. Hence, after a major solar flare, airline companies have to re-route these flights—pushing up the operating costs in time and aviation fuel. Nowadays we have many man-made satellites going around the earth and serving diverse purposes from broadcasting television programmes to providing signals to the global positioning system (GPS). A big flare can roast the electronics on-board a satellite unless expensive precautions are taken.

Interestingly, the first flare recorded by any human being—Carrington's 1859 flare—seems also to be the strongest flare so far recorded by any human being. This flare caused magnetic storms as far away from the geomagnetic pole as India and was recorded by the instruments in the Indian Institute of Geomagnetism in Mumbai. In 2003 Bruce Tsurutani, a Japanese-American scientist who was tracking the effects of this flare, and Gurbax Lakhina, the then Director of the Indian Institute of Geomagnetism, unearthed these old records. They concluded that this flare was at least three times as strong as the 1989 flare which caused the blackout in Quebec![2] The world was technologically much less advanced in 1859. The most advanced electrical technology used at that time was the telegraph. Although extraordinary reports of unusually brilliant auroral displays came from around the world and many telegraph lines went dead, lives of people were not paralyzed like the lives of Quebec citizens in the morning of 13 March 1989. If a flare of strength comparable to that of the 1859 flare were to unleash its full fury on the earth today, it would have a catastrophic effect on our present-day technology-dominated world, probably causing various kinds of damage all over the world running into billions of dollars. Ironically, as we are becoming technologically more advanced—as we have large national power grids and electronic hardware in space, depending more and more on communication through satellites—we are also becoming more vulnerable to solar disturbances.

It is not at all easy to see from the basic principles of physics why an explosion on the sun would cause such things as a sudden change in the earth's magnetic field. In the closing decade of the nineteenth century, as more evidence started coming up to show a connection between solar disturbances and geomagnetic storms, the ageing Lord Kelvin, one of the creators of modern thermodynamics and the most celebrated living physicist of Britain at that time, was completely flummoxed by claims of such a connection between solar disturbances and geomagnetic storms. He was convinced that basic principles of physics do not allow such a connection and that the people who were making such connections were gravely in error. As the President of the prestigious Royal Society, Lord Kelvin launched a broadside in his presidential address of 1892. A few years earlier, James Clerk Maxwell—who was arguably the greatest theoretical physicist in the intervening period between Newton and Einstein, and who had already died at the relatively young age of 48—had shown that all the basic principles of electromagnetic theory can be put in the form of a set of elegant equations. We shall have to come back to Maxwell's equations on several occasions later in the book. Lord Kelvin used Maxwell's equations to demonstrate that even a very violent explosion on the sun would not produce any appreciable change in the earth's magnetic field.

Lord Kelvin certainly started with the correct equations and made no mistakes in his calculations. However, one assumption he had made was that the space between the sun and the earth was completely empty—an assumption which nobody questioned for several decades until 1958 when a young man of 31 named Eugene Parker proposed a most extraordinary theory. It was known by that time that the sun's corona—the beautiful halo around the sun seen during a total eclipse—was much hotter than its surface. Working with the basic equations of fluid mechanics, Parker found that the high temperature of the corona should cause a continuous wind of hot gas starting from the sun and blowing through the entire solar system. Within a few years, measurements from satellites brilliantly confirmed Parker's theory. Parker had named this *solar wind*. The existence of this wind changes things completely and Lord Kelvin's arguments no longer go through if this wind is present. We now know that the sun can affect the earth in very complicated ways which nobody could suspect as long as the space between the sun and the earth was thought to be empty. While light from the sun takes about eight minutes to reach the earth, the solar wind takes about three days to cover the sun–earth distance and that is often how long it takes a solar flare also to affect the earth. This suggests the tantalizing possibility that the debris of a solar flare may be carried to the earth by the solar wind. We shall discuss this in more detail in later chapters.

1.2 A Recurring Cosmic Illness

A flare is certainly a symptom that the sun is stricken with some kind of cosmic 'illness'. However, if we want to use flares as a diagnostic tool to assess the sun's health, we run into one problem. A flare is a very temporary event lasting for a few minutes. Even when the sun is stricken with this mysterious illness, a flare may not occur precisely at the moment that you have chosen to look at the sun. However, we have seen that flares usually occur above large sunspots and a large sunspot typically lasts for many days. The best way of figuring out when the sun is stricken with this cosmic illness is to monitor the number of sunspots seen on the sun's surface.

You do not need a telescope to see large sunspots. They are visible to the naked eye, when the sun is near the horizon just after sunrise or before sunset. Aristotle had taught that the sun was a perfect sphere of fire without any blemishes, and the medieval church considered it to be close to a blasphemy to talk about blemishes on the sun. However, a large sunspot was seen before Emperor Charlemagne's death in 814 and some medieval chroniclers considered it as an omen foretelling the imperial death. Galileo and some of his contemporaries during the Renaissance were the first Europeans to study sunspots systematically and to leave records behind them. But sunspots have been studied in some oriental countries for nearly 2000 years. In ancient China and Korea, it was believed *officially* that sunspots foretell the emperor's fortune—exactly what Charlemagne's ministers suspected. So the emperors had their official sunspot watchers. Life for the world's first experts on sunspots was not always easy. If a sunspot watcher could not correctly interpret the message a sunspot had for the emperor, then he would be punished severely—sometimes possibly even with death. Luckily present-day experts on sunspots like me do not have to face such serious consequences when we occasionally make wrong predictions. As a colleague once remarked, if the ancient tradition of harsh punishments still continued, it is possible that many bad scientific papers would not get written and that might have a beneficial effect on the health of our journals. To be fair to the oriental scholars, we have to admit that they were not altogether wrong in assuming that sunspots had effects on the lives of their emperors. We, of course, have very few emperors left in today's world. However, as we now realize, sunspots can be the cause

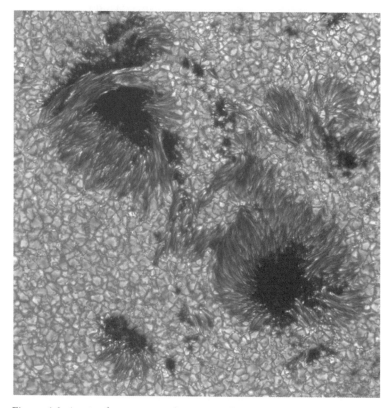

Figure 1.3 A pair of sunspots on the sun's surface. We shall discuss in the next chapter why sunspots very often appear in pairs. This image was taken with the Swedish 1-m Solar Tower Telescope in La Palma island on 14 August 2003. Credit: G. Scharmer, K. Langhans and M. Löfdahl, Institute for Solar Physics, Sweden.

of solar explosions which affect lives of not only emperors, but lives of commoners like you and me as well.

The sun seems to get a particularly severe attack of sunspots every 11 years, when the heavenly face of the sun gets pockmarked with many of them. In 1844, a few years before Carrington discovered the solar flare, Heinrich Schwabe, an apothecary in the small German town of Dessau, discovered the sunspot cycle. Astronomy was Schwabe's passion. After procuring a small telescope in 1826, he systematically observed the sun every day whenever the sky was clear. His aim was to

Figure 1.4 Heinrich Schwabe (1789–1875), the discoverer of the 11-year sunspot cycle.

discover any planet closer to the sun than Mercury. If such a planet existed, then Schwabe expected this planet to sometimes come in front of the sun, making it visible as a dark spot on the disc of the sun. Such transits of both Mercury and Venus across the disc of the sun have been seen. While searching for this planet, Schwabe kept a record of the sunspots which he saw. After he began observations in 1826, he saw many sunspots in the first few years. He counted as many as 225 in 1828. Then he seemed to find fewer sunspots in the next few years, counting only 33 in 1833. After this, Schwabe again started to see more sunspots in subsequent years. The number then started decreasing again. By 1844 it became clear to Schwabe that he had discovered a periodic cycle in the appearance of sunspots. Announcing his discovery, he wrote:

> The weather throughout this year was so extremely favourable that I have been able to observe the Sun clearly on 312 days; however, I counted only 34 groups of sunspots... From my earlier observations, which I have

reported every year in this journal, it appears that there is a certain periodicity in the appearance of sunspots and this theory seems more and
more probable from the results of this year.[3]

Schwabe estimated the period of this cycle to be about 10 years. Now we
know that it is closer to 11 years. Not all cycles have the same length.
Some cycles can be a little bit longer than 11 years and some can be
shorter.

Figure 1.5 shows how the number of sunspots has varied with time.
How to standardize sunspot number counts is a technical subject which
we shall not get into. The vertical axis of Figure 1.5 is based on a
commonly used modern standardization. A few years after Schwabe's
discovery of the sunspot cycle, Rudolf Wolf, the Director of Bern Observatory, started a systematic record of sunspots involving several
observatories. Nobody had bothered to keep such systematic records
of sunspots before about 1850, and only sporadic records exist, depending on whether a particular astronomer who happened to look at the
sun on a particular day left a record of the sunspots seen on that day
in the ledger of some observatory or in the pages of a published book.
A few years ago, an American scientist named John Eddy delved into
the records of several old observatories and estimated the numbers of
sunspots which must have been seen in different years before systematic recording began. The first entries in Figure 1.5 a little after 1600
are based on the records left by Galileo and his contemporaries. Then
we find a period from about 1640 to 1720 when sunspots were rarely
seen. Eddy named this period the 'Maunder minimum'—after Edward
Maunder who was one of the first scientists to point out, towards the

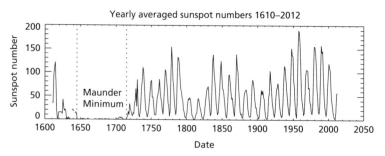

Figure 1.5 The yearly averaged number of sunspots plotted against time for
the period 1610–2012. Credit: David Hathaway.

end of the nineteenth century, this unusual seventeenth century epoch when sunspots disappeared. After the Maunder minimum, the number of sunspots seems to be going up and down in a periodic fashion. For convenience, scientists have labelled the cycles with a number starting from the late eighteenth century. The latest cycle which started in 2008 is numbered 24. It is clear that not all cycles have been equally strong. Some cycles have been fairly weak with few sunspots even during their peaks. Others have been much stronger.

If solar disturbances like flares are linked to sunspots, then we should expect to have more of them when the sun has more sunspots. This turns out to be precisely the case. The occurrence of solar disturbances becomes more frequent when a sunspot cycle reaches its peak. Because of their many consequences on our present-day technology-dependent society, we have to get geared up for solar disturbances as a sunspot cycle approaches its peak. Also, it is seen that stronger sunspot cycles are associated with larger numbers of solar disturbances. However, there is no evidence that individual solar disturbances of a stronger cycle are on average any stronger than the individual solar disturbances of a weaker cycle. A stronger cycle simply produces more of them. One of the hot questions of present-day research is whether we can predict the strength of a future sunspot cycle before its onset and thereby be prepared for what lies ahead. Apart from sunspots and flares, many other things connected with the sun vary in tandem with the cycle. For example, the appearance of the sun's corona changes dramatically with the sunspot cycle. Hence the sunspot cycle is often simply called the *solar cycle*. We say that the sun is more *active* during the peak of the cycle. Sunspots and flares are among the most easily visible manifestations of solar activity.

1.3 Nature's Third Cycle

There are certain cyclic phenomena in nature which affect our lives in very big ways. The most important cycles, of course, are the day-night cycle and the cycle of seasons. Even the lowliest of animals are aware of the day-night cycle. Human civilization can be said to have begun when our ancestors formed a clear grasp of the cycle of seasons. They had to know when to sow seeds and when to do the harvesting. It was this knowledge that made the first settled life possible. There was another great cycle of nature—the cycle of the moon's phases—which was very important for primitive man. On a full moon night, he could see his surroundings and defend himself against predatory animals. In some of the world's major religions—such as Islam and Hinduism—important festivals still follow the lunar calendar. Chinua Achebe, in his powerful novel *Things Fall Apart* depicting the pre-colonial tribal life in Africa, wrote:

> The night ... was always quiet except on moonlight nights. Darkness held a vague terror for these people, even the bravest among them. Children were warned not to whistle at night for fear of evil spirits. Dangerous animals became even more sinister and uncanny in the dark...On a moonlight night it would be different. The happy voices of children playing in open fields would then be heard. And perhaps not so young would be playing in pairs in less open places, and old men and women would remember their youth.

However, for most of us living in big cities, the cycle of the moon's phases has ceased to have any relevance to our daily lives. If you ask an average city-dweller what the phase of the moon is today, you are not likely to get an answer in most cases. As our society has become more technologically advanced, the cycle of the moon's phases has lost its importance. At the same time, the sunspot cycle has gained in importance as we have become more dependent on technology. I would argue that we have already reached that phase of civilization where the sunspot cycle has replaced the cycle of the moon's phases as nature's third most important cycle for our lives. The aim of this book is to tell readers about our current understanding of this third of nature's most important cycles.

Sir Francis Bacon was a remarkable scholar-statesman in Renaissance England. There are literary scholars who believe that he penned all the 37 plays normally ascribed to a man named William Shakespeare! While others may find it hard to swallow this theory, Bacon was an eloquent champion of science. He argued that scientific discoveries would lead to the betterment of people's lives—a rather novel and unconventional idea at that time. Bacon proposed a methodology for how science should be done. His idea was that the job of scientists is to collect experimental data in a completely unbiased manner, without being prejudiced by theoretical notions in any way. Later philosophers of science have criticized Bacon for proposing something which clearly appears a naive and faulty model of science. A mature branch of science usually has some central theoretical paradigms. A scientist is usually guided by these central paradigms while collecting and organizing data. While the Baconian model may not be a good model for science in general, this model perfectly describes the nature of research on solar activity for about a century from the time of Schwabe's 1844 discovery of the 11-year sunspot cycle. Solar astronomers gathered more and more data about various aspects of the sunspot cycle. But there were no central ideas around which the data could be organized systematically. Nobody had a clue about how to explain sunspots and solar flares or why there is a 11-year cycle. The study of solar disturbances and the sunspot cycle was a purely Baconian enterprise for about a century.

It may be mentioned in passing that in 1795 William Herschel, the discoverer of the planet Uranus and perhaps the greatest astronomer of his age, made an attempt to provide an explanation of sunspots. He believed that the sun had a cool surface surrounded by hot radiating clouds. The clouds were so densely packed in the solar atmosphere that we usually saw only the tops of these hot radiating clouds. Only occasionally did gaps in the clouds reveal the cooler interior. According to Herschel, sunspots were just these gaps between the radiating clouds.[4] He even conjectured that there might be living beings in the cooler interior of the sun! This theory of Herschel should caution us that in science we should not believe everything that even a very great man says. However, do not conclude that this theory of Herschel's has been irretrievably consigned to the dustbin of history. Discarded scientific theories sometimes have an almost uncanny capacity for reincarnating themselves in new avatars. Nineteenth century physicists trashed Newton's corpuscular theory of light in favour of the wave theory. But, in less than a century, evidence started mounting that, under certain circumstances, light actually behaves like a flux of corpuscles (which we now call photons). About 1940 solar astronomers started getting the first indications that certain aspects of Herschel's theory may be correct after all. Astronomers, for sure, did not discover a cooler interior of the sun teeming with living creatures like us. But what they discovered was hardly any less incredible. One implication of Herschel's theory was that the temperature should increase as we go above the cool surface of the sun. Astronomers started getting irrefutable evidence that this was indeed the case. The temperature of the solar surface is about 6000 K. As we go above this surface, the temperature keeps increasing until it reaches a few million degrees in some parts of the corona—being only slightly less hot than the central core of the sun. Later in the book we shall have occasion to discuss the mysteriously hot corona of the sun. Especially, when we address the question of how a flare can affect the earth, we shall see that the hot corona will provide the crucial missing link in our chain of arguments. I have already mentioned that the hot corona drives the solar wind from the sun.

1.4 The Magnetic Breakthrough

A really major breakthrough in our physical understanding of what sunspots actually are came about a century ago when George Ellery Hale discovered in 1908 that sunspots are regions of strong magnetic field. Let us go back in time a little bit to discuss the role of magnetic fields in astronomy. Around 1600 William Gilbert, physician to Queen Elizabeth I of England, proposed a bold hypothesis to explain why a suspended compass needle points in the north-south direction. He suggested that the whole earth is a huge magnet and attracts the compass needle. This was probably the first time that somebody proposed an astronomical object—the planet earth—to have a large-scale magnetic field. Initially it was thought that the earth was a gigantic lodestone. Lodestones contain ferromagnetic substances like iron which can remain magnetic even when left to themselves. However, by the end of the nineteenth century, it became clear that a ferromagnetic substance loses its magnetism when heated beyond a certain temperature known as the Curie point. From volcanic eruptions and other considerations, it was clear that the interior of the earth is hotter than the Curie temperature of any known ferromagnetic substance. It became apparent that the earth could not be a gigantic lodestone and one has to look for alternative explanations for the earth's magnetic field.

Astronomers since Gilbert's time have also bothered about the question whether the earth is a unique object in being magnetic or whether other astronomical bodies also have magnetic fields. The difficulty in answering this question was that nobody knew till the beginning of the twentieth century how to determine if an astronomical body has a magnetic field. The easiest way of checking whether there is a magnetic field somewhere is to take a small compass needle there and to see whether it is deflected. Certainly one could not take a compass needle to the sun to look for the sun's magnetic field. In 1897 Pieter Zeeman made an important discovery, for which he won the Nobel

Figure 1.6 George Ellery Hale (1868–1938), the discoverer of magnetic fields in sunspots.

Prize five years later and which finally suggested a way of looking for magnetic fields in astronomical objects. On keeping a source of light in a strong magnetic field, Zeeman found that there were multiple lines in the light spectrum where there would have been a single line in the absence of the magnetic field. Barely a decade after Zeeman's discovery, Hale noted in 1908 that the spectrum of a large sunspot showed this splitting of spectral lines. He concluded from this that a sunspot is a region of strong magnetic field.[5] The magnetic field of a large sunspot can be about 5000 times stronger than the magnetic field around the geomagnetic pole on the earth's surface. For readers familiar with the unit *tesla* for magnetic field, the typical magnetic field in the interior of a large sunspot is about 0.3 tesla. This is only a little weaker than magnetic fields produced by powerful electromagnets, but very strong by human standards. If you have an electromagnet producing a magnetic field of 0.3 tesla, you will not be able to hold a small piece of iron like a nut or bolt in your hand in that magnetic field. It will get wrenched out of your hand and get slammed into the nearby pole of the electromagnet. Hale's discovery of magnetic fields in sunspots was truly a momentous discovery in the history of physics because this was the first time that somebody conclusively established the existence of magnetic fields outside the earth's environment. One of the major developments in twentieth century astronomy is the realization that magnetic fields are ubiquitous in the astronomical universe. Most stars, planets and galaxies have magnetic fields. The study of magnetic fields in astronomy has blossomed into a major research field.

With Hale's discovery, it became clear that the 11-year sunspot cycle is nothing but a magnetic cycle of the sun. Since solar flares usually occur above large sunspots, it also seemed probable that solar flares must have something to do with magnetic fields. That gave a first clue to addressing the question of why sudden changes in the earth's magnetic field are observed after major solar flares. But, to begin with, why does the sun have a magnetic cycle?

In high school we all learn that matter can exist in three states: solid, liquid and gas. Physicists, however, have been aware for about a century now that there can be a fourth state of matter: the plasma state. We know that the atom is not the indivisible fundamental building block of matter which it was once supposed to be. We have negatively charged electrons inside the atom going around a positively charged nucleus at the centre. When a gas is heated to a high temperature, electrons start getting knocked out from the atom leaving the positively charged remainder—called the ion—behind. This process—known as *thermal ionization*—was first studied systematically by the Indian physicist Meghnad Saha in 1920. The collection of electrons and ions is called the plasma. Saha's theory tells us that matter inside and around the sun must exist in the plasma state because of the high temperature. In fact, it is estimated that more than 99% of the visible matter in the astronomical universe is in the plasma state. The earth, which is our home, is a truly unusual place in the universe where the plasma state is not the most abundant state of matter.

It has been known from theoretical considerations and from laboratory experiments that a moving plasma can interact with a magnetic field in complicated ways. The branch of physics in which we study how magnetic fields behave inside a moving plasma is known by the horrible tongue-twisting name 'magnetohydrodynamics'—usually shortened to MHD, where M stands for 'magneto-', H for '-hydro-' and D for '-dynamics'. The first steps for developing this subject were taken around 1930–1950. One of the central questions in the theory of MHD is the following: is it possible for the plasma inside an astronomical body to move in such a fashion that the astronomical body continues to have a magnetic field as long as the plasma moves in this way? A system which continues to have a magnetic field because of the movements of plasma inside it is called a *self-excited fluid dynamo*. The crucial question is whether basic equations of MHD allow the existence of such self-excited fluid dynamos. If self-excited fluid dynamos do exist, then one could hope to explain the magnetic fields of the earth, the sun and other astronomical systems simply by suggesting that all these objects have self-excited fluid dynamos inside them. The theoretical study of such self-excited fluid dynamos is called *dynamo theory*.

The first path-breaking result in this field, however, was a rather disappointing negative result. When dealing with complex equations, one of the first things that a smart physicist tries to do is to find out whether the equations admit of a simple solution. That is what Thomas George Cowling did with the MHD equations when he tried to solve the self-excited dynamo problem by considering sufficiently simple and symmetric plasma motions. In 1933 he announced a famous anti-dynamo theorem that MHD equations do not allow the possibility of a self-excited fluid dynamo if you restrict yourself to sufficiently simple symmetric plasma motions inside your system.[6]

Cowling's anti-dynamo theorem naturally caused some amount of consternation in the field. There arose two schools of thought. The first school of thought was that Cowling's theorem is a special form of a more general theorem that self-excited fluid dynamos are completely impossible with any kind of plasma motion. The second school of thought was that we must be able to demonstrate the existence of self-excited fluid dynamos by considering sufficiently complex unsymmetric plasma movements. One of the prominent proponents of the second school of thought was Walter Elsasser. For many years he tried to solve MHD equations with various kinds of complex plasma movements. But he failed to demonstrate the existence of a self-excited fluid dynamo even after struggling for several years. Fritz Krause, a distinguished German dynamo theorist, has given the following interesting account of an encounter between Elsasser and his friend Einstein:

> Walter Elsasser and Einstein were friends in Germany before they both emigrated to the US in the 1930s. Several years after Elsasser had settled there (in the late 1930s in fact), he became interested in the origin of the geomagnetic field. Einstein paid him a visit, and (as people do) asked 'What are you working on these days?' Elsasser told him, and Einstein

invited him to explain dynamo theory to him. Elsasser set up the problem and then told Einstein about Cowling's theorem. Einstein's response
was, 'If such simple solutions are impossible, self-excited fluid dynamos
cannot exist.' For once, the great man's craving for simplicity seems to
have misled him.[7]

At this critical juncture, there appeared on the scene a young man
whom we have already mentioned for his theory of the solar wind and
whom we shall repeatedly meet in the pages of this book. The full name
of this young man was Eugene Newman Parker—affectionately known
as Gene in the solar physics community. As a research associate of Elsasser, Parker began his meteoric career with an astounding feat. In 1955,
at the age of 28, he solved the self-excited fluid dynamo problem by
showing for the first time that MHD equations allow the existence of
a self-excited fluid dynamo. What is more, he derived the fundamental equation in this field known as the *dynamo equation* and, on solving
it, he found that he could explain many aspects of the sun's magnetic
field, including the 11-year cycle which had appeared so completely
mystifying to everyone until that time.

Figure 1.7 Eugene Newman Parker (1927–), photographed by the author
when he was Parker's PhD student.

You may think that a young man who solves a long-standing scientific problem in this way would immediately be hailed as a hero of science. But things did not happen quite that way. Parker's extraordinary paper on the self-excited dynamo problem, which also provided the first theoretical explanation of the 11-year cycle, appeared in *Astrophysical Journal* in 1955.[8] It was a moderately long paper of 22 pages bristling with unfamiliar and formidable-looking equations. Some of the persons who had earlier attempted and failed to solve the self-excited dynamo problem included Cowling, Alfvén, Elsasser and Chandrasekhar. They were all exceptionally capable scientists and Parker probably guessed that, if it was possible to crack the dynamo problem by the methodologies which they were following, one of them would have surely cracked the problem. So Parker had to introduce some radically new ideas and a totally unorthodox methodology to solve the dynamo problem. We shall discuss some of these new ideas later in the book. The result of all these was that most solar astronomers found Parker's paper on the dynamo completely incomprehensible to them. Probably very few people read the paper at the time of its publication and some who read failed to grasp its significance. What was personally worse for Parker, the authorities of the University of Utah where he was employed at that time did not seem interested in regularizing his job in spite of this work. Apart from Elsasser, one other person whom Parker had thanked in his dynamo paper for encouragement was Subrahmanyan Chandrasekhar, regarded by many as the greatest theoretical astrophysicist the world has so far seen. Luckily, the University of Chicago where Chandrasekhar was working agreed to offer Parker a job. Parker moved there and spent the rest of his career at the University of Chicago.

People often ask me whether we now fully understand the reason behind the 11-year sunspot cycle. People who ask this question usually expect a simple answer like 'yes' or 'no'. As it happens, the answer to this apparently straightforward question is reasonably complex. There are some scientific questions to which we have fairly definitive answers. After Newton formulated the theory of universal gravitation and the basic laws of motion, the problem of planetary motion found a nearly complete solution in the hands of Newton. In the twentieth century, the structure of the hydrogen atom was understood completely by solving Schrödinger's equation, the fundamental equation of quantum mechanics. The planetary system and the hydrogen atom are, in the views of physicists, particularly simple and clean systems. Usually complete understanding of a physical system is possible only if the system is simple and clean like this. That is not the case for the sunspot cycle. This cycle is caused by the combined effect of many complicated processes. Our understanding of such systems usually advances by many incremental steps. Sometimes there are major breakthroughs, like Parker's 1955 paper on the fluid dynamo. But even this paper provided a guiding beacon and a blueprint for future development rather than a full solution. Many of us, guided by the deep insight provided by Parker's work, have attempted to work out detailed theoretical explanations of different aspects of the sunspot cycle. It seems that we have gradually reached the stage where there is more light than darkness— at least that is the view which some of us hold now. But we shall probably never understand the sunspot cycle the way we understand the planetary system or the hydrogen atom. One of the aims of this book is to tell the story of our present understanding of the sunspot cycle. Even in its incomplete and unfinished state, it is a grand story worth telling.

My own association with the sunspot cycle began in the early 1980s when I had the privilege of working as a PhD student at the University of Chicago under the supervision of Gene Parker. Several of Parker's students are now leading theoretical solar scientists. However, amongst Parker's students who have turned out to be successful scientists, with the exception of me, no one else has worked on dynamo theory in the past few years. If I have any claim to fame, it must be on the ground that I am the only working dynamo theorist in the world who was trained by the Master himself.

1.5 A Timeline

I end with a timeline of important landmarks in this field which have been mentioned in this chapter though not in a chronological order. Now these landmarks are listed chronologically.

- 1600: Gilbert proposed that the earth is a magnet.
- 1610: Galileo began his study of sunspots using the telescope.
- 1844: Schwabe discovered the sunspot cycle.
- 1859: Carrington observed a solar flare (the first such observation by anybody) and noted the geomagnetic storm several hours later.

- 1892: Kelvin 'proved' that solar disturbances cannot affect the earth.
- 1908: Hale discovered that sunspots have strong magnetic fields.
- 1933: Cowling proved the anti-dynamo theorem.
- 1955: Parker formulated the dynamo equation and proposed the first theoretical model of the sunspot cycle.
- 1958: Parker developed the theory of the solar wind, which was discovered a few years later.

2

The Mysterious Sunspots

2.1 Sunspots Reveal the Sun's Rotation

Like many other things in modern science, solar physics—the scientific study of the sun based on the principles of physics—can be said to begin with Galileo. Galileo lived in an age when priority disputes used to be a common part of a scientist's life. Today a scientist usually reports a new discovery in the pages of a scientific journal and its publication date provides irrefutable proof of when the discovery was made. But, in Galileo's time, there was no standardized procedure for dissemination of new scientific discoveries. Sometimes extra complications arose due to extraneous factors. For example, Christoph Scheiner, who was a contemporary of Galileo and made telescopic observations of sunspots at the same time, was a Jesuit priest and needed permission from the Jesuit authorities to publish the results of his scientific investigations. Theodore Busaeus, his superior in the Jesuit hierarchy, forbade him to publish his results on sunspots with the following admonition:

> I have read my Aristotle from end to end many times and I can assure you that I have never found in it anything similar to what you mention. Go, my son, calm yourself, and be assured that what you take for spots on the sun are the faults of your glasses or your eyes.[1]

All Scheiner could do at that time was to publish a short pamphlet about sunspots secretly under a pseudonym.

Around 1610 four men began telescopic observations of sunspots, approximately at the same time—Galileo Galilei, Christoph Scheiner, Johannes Fabricius and Thomas Harriot. A fierce dispute raged amongst some of them as to who was the first. This was a question which certainly could not be settled using records publicly available then or now. I shall refrain from getting into the details of this unfortunate priority dispute and take the simple point of view that all four of them deserve credit as pioneers. I recommend Judit Brody's wonderful book *The Enigma of Sunspots* to readers desirous of learning more about the pioneers of sunspot research. Since Galileo had the sharpest mind amongst the four pioneers, he succeeded in drawing more far-reaching conclusions from sunspot observations than the others.

It was found that a sunspot keeps changing its position on the sun's disc from day to day. Galileo drew the correct conclusion that sunspots must be marks on the solar surface and that they seem to change their positions because the sun is rotating about its own axis, just like the earth. From the shifts in the positions of sunspots from day to day, Galileo estimated that the rotation period of the sun is about 27 days. I would say that this discovery marks the true beginning of solar physics. This is the first time that somebody used a good logical argument to draw a tremendously important conclusion about a physical condition of the sun. As we shall see, the rotation of the sun plays a central part in our story of how the 11-year sunspot cycle is produced.

We all know that the earth rotates around its axis in 24 hours, causing day and night. The earth rotates like a solid body, which means that each and every part of the earth takes the same time of 24 hours to go once around the rotation axis. Astronomers initially expected the sun also to rotate like a solid body. Richard Carrington, whose acquaintance we have made for his observation of a solar flare, was the first to realize that this may not be the case. By the middle of the nineteenth century, telescopes had improved tremendously compared to Galileo's time and several astronomers were engaged in a more accurate determination of the sun's rotation. Very surprisingly, the results of these much more accurate measurements, instead of pinning down the value of the sun's rotation period more accurately, seemed to produce more errors and uncertainties. Finally Carrington solved the puzzle in 1859 when he realized that all the sunspots do not go around the rotation axis at the same rate.[2] A sunspot near the sun's equator would take about 25 days to go around the sun's rotation axis, whereas a sunspot at a higher latitude would take a longer time. I may mention that a sunspot typically lasts only for a few days. But, from measurements of the daily positions of a sunspot, one can figure out how long the sunspot would take to go around the rotation axis if it lasted for many days. Although we do not see sunspots near the sun's poles, it has been found by other methods that a portion of the sun near the pole takes about 35 days to go around the rotation axis.

In astronomical jargon, this kind of rotation is called *differential rotation*. Regions near the sun's equator move faster around the rotation axis. We can try to imagine what would happen if the earth suddenly started having differential rotation. I live in Bangalore, which more or less lies straight below New Delhi in a map in which south is taken in the downward direction. If the regions near the earth's equator suddenly started moving faster around the rotation axis than the regions at higher latitudes (as in the case of the sun), then after some time, Bangalore would be down below Kolkata and then eventually down below Beijing or Tokyo! Luckily such things do not happen. If the earth had differential rotation, then all the continents would be distorted continuously and a world map would have to be labelled with the time of its validity. There are no continents on the surface of the sun that could be distorted. But the surface of the sun is not entirely featureless when groups of sunspots are present. We have discussed Hale's discovery of the strong magnetic field in sunspots. The differential rotation of the sun keeps distorting the magnetic field of the sun. We shall see later that this distortion of the sun's magnetic field by differential rotation is a key ingredient in developing a dynamo model to explain the 11-year sunspot cycle.

2.2 The Little Ice Age

As I have pointed out in Section 1.2 and as can be seen in Figure 1.5, there were no sunspots seen during the epoch 1640–1720. John Eddy, who in 1976 published a landmark study of this unusual epoch, named it the 'Maunder minimum'.[3] Edward Maunder, along with Gustav Spörer, in the 1890s tried to draw people's attention to the absence of sunspot records during the epoch 1640–1720. But their writings on this subject seemed to have been totally ignored. Finally Eddy succeeded in marshalling different kinds of evidence to show that something unusual was indeed happening during this epoch and at last forced the scientific community to take the matter seriously. Eddy pointed out that this was precisely the epoch during which reports of auroral sightings were also rare and astronomers who observed total solar eclipses during this epoch did not mention the corona at all—probably because the corona also had gone missing during this epoch.

Since there were very few astronomers studying sunspots systematically during this epoch, there was one obvious reason why the earlier writings of Maunder and Spörer were not taken seriously. As Eddy had put it, can the 'absence of evidence' be taken as 'evidence of absence'? In other words, could the absence of sunspot recordings during this epoch simply be due to the fact that there were no astronomers systematically looking for sunspots? This question has by now been settled beyond any reasonable doubt. There were indeed astronomers during this epoch who were aware of earlier observations of sunspots and were looking for them, but only occasionally would see a rare sunspot. Giovanni Cassini, the Italian-born astronomer who worked in the Paris Observatory for many years and is remembered mainly for his work on the gaps between Saturn's rings, saw one such rare sunspot in 1671 and wrote:

> it is now 20 years since astronomers have seen any considerable spots on the sun, though before that time, since the invention of the telescopes they have from time to time observed them.

The Maunder minimum was undoubtedly a reality.

There are some modern scholars who go to the opposite extreme of arguing that many astronomers were regularly looking for sunspots since the invention of the telescope and the 11-year cycle would have

Figure 2.1 Edward Walter Maunder (1851–1928), a pioneering solar physicist of his time.

been discovered much before Schwabe's discovery in 1844 if the Maunder minimum had not taken place. I find it difficult to accept this view. The sun came out of the Maunder minimum around 1720 and the 11-year cycle was discovered by Schwabe more than a century later based on about 17 years' careful observations. Had there been a meticulous sunspot watcher like Schwabe when the sun came out of the Maunder minimum, the 11-year cycle could have been discovered by 1740. While the Maunder minimum was real, there is no doubt that no astronomer bothered to keep a regular record of sunspot sightings for a long time. In fact, William Herschel (whose theory of the sun's cool surface has been mentioned in Section 1.3) was fairly close to discovering the sunspot cycle about half a century before Schwabe. He noticed that many sunspots were seen in some years and very few in other years. He suspected that there might be a periodicity, but he did not have reliable good data extending over several years to draw any firm conclusions. Agnes Clerke, the extremely talented popularizer of astronomy in Victorian England, presumably got the correct reason behind this delay in the discovery of the sunspot cycle when she wrote:

> The changes visible in the solar surface were then generally regarded as no less capricious than the changes in the skies of our temperate regions. Consequently, the reckoning and registering of sun-spots was a task hardly more inviting to an astronomer than the reckoning and registering of summer clouds.[4]

Eddy pointed out another very curious thing. The Maunder minimum seemed to coincide with what is often called the little ice age—a period of the late seventeenth century when much of Europe had unusually cold winters and peasants had to go through great hardship. Although this was before regular record-keeping of weather had begun, we have some pictorial evidence of these unusual winters. The seventeenth century was the golden age of Dutch painting (the age of Rembrandt) when the new genre of landscape painting was becoming popular. Some Dutch painters have left behind winter landscapes of frozen rivers and canals which never freeze in winters nowadays. The river Thames near London also often froze during winters in this epoch. Was it merely coincidence that the disappearance of sunspots and this unusual cold climate occurred at the same time? Or did the missing sunspots have something to do with the little ice age? Later in the book I shall present more evidence to support the view that the earth indeed

Figure 2.2 The painting of a frozen river in winter by the Dutch landscape painter Aert van der Neer, who painted landscapes during the Maunder minimum.

becomes cooler when sunspots go missing. Exactly how this happens is still a question on which experts seem to have very differing views and which is unlikely to be settled definitively in the near future.

An epoch like the Maunder minimum is called a *grand minimum* of the sun, to distinguish it from the ordinary sunspot minimum between two sunspot cycles when very few sunspots are seen for a couple of years. Have there been other grand minima in the past before the discovery of the telescope? In Chapter 9, I shall discuss certain techniques by which this question can be answered. The answer is in the affirmative: there have been several such grand minima in the last few millennia. Of late, solar astronomers have also started talking about grand maxima—epochs during which several successive sunspot cycles have been unusually strong. Another look at Figure 1.5 will convince you that much of the twentieth century has been an epoch of grand maximum. If a grand minimum produces a cooling of the earth, will a grand maximum like the one in the twentieth century lead to a heating of the earth? Unfortunately this question has not remained a purely scientific question and has been badly mired in political controversy. The twentieth century has been a period of global warming—presumably due to man-made greenhouse gases. People who want to deny this have found a convenient alibi in the sun, blaming the increased number of sunspots in the twentieth century for global warming. It is becoming increasingly difficult to separate sound science from political opinion in this debate. We shall discuss this more in Chapter 9.

William Herschel claimed that the wheat price in Britain varied with the occurrence of sunspots. Over the years, there have been many attempts to correlate various things with the sunspot cycle or the Maunder minimum. It is sometimes extremely difficult to ascertain the amount of truth in these claims. Violins made by Antonio Stradivari have never been surpassed in musical quality. While he may have been an exceptionally skilled violin-maker, it has been pointed out that he lived exactly during the Maunder minimum and the spruce wood used for making violins in that period had some special qualities due to the unusual winter conditions. It is difficult to prove or disprove such theories, but you may want to believe the Stradivari story because it decidedly makes an interesting conversation piece at a party and, by telling your friends this story, you can show that you are very cultured.

2.3 The Solar Butterflies

Since the sun rotates about a rotation axis like the earth (although it does have differential rotation unlike the earth), we can introduce poles and the equator with respect to this rotation axis just as in the case of the earth. Even the early pioneers like Galileo noticed that sunspots do not appear all over the sun. They are primarily seen in two belts on the two sides of the equator. A few years after Schwabe's discovery of the sunspot cycle, Carrington discovered what is called the latitudinal drift of sunspots.[5] He found that sunspots at the beginning of a cycle appear around latitudes of 30° on both sides of the equator. Then, as time goes on, the newer sunspots of the cycle appear progressively at lower and lower latitudes. Eventually, at the end of the cycle, sunspots are seen fairly close to the equator. After that, the next cycle begins with sunspots again appearing around 30° of latitude.

In 1904 Maunder (after whom the Maunder minimum is named) plotted these observations in an instructive diagram, with time along the horizontal axis and the solar latitude along the vertical axis.[6] Figure 2.3 shows one of Maunder's original diagrams. At a particular time, the latitudes where sunspots are seen are shaded. For example, you can see in Figure 2.3 that the year 1890 was the beginning of a cycle when sunspots appeared at high latitudes on both sides of the equator, as indicated by the shading of these latitudes above the 1890 mark on the horizontal axis. As you move towards the right corresponding to later times, lower latitudes on both sides of the equator are shaded,

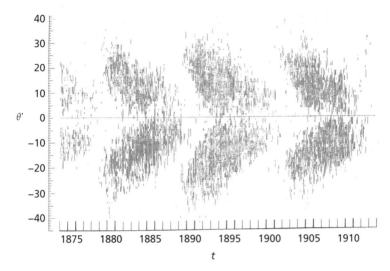

Figure 2.3 A butterfly diagram constructed by Edward Maunder. Such diagrams show the latitudes in which sunspots had been seen at different times.

corresponding to sunspots appearing progressively at lower latitudes. Since the resulting diagram reminds one of a repeated pattern of butterflies, it is called the *butterfly diagram*.

At exactly the same time when Carrington was studying the latitudinal drift of sunspots with the cycle, the German solar astronomer Gustav Spörer also noticed it and started studying it systematically.[7] As it happened, Carrington could not carry out his studies for too long due to problems in his personal life. Spörer studied the latitudinal drift of sunspots with full thoroughness and the movement of sunspot belts on both sides of the equator towards lower latitudes with the progress of the sunspot cycle is now known as *Spörer's law*.

We have already discussed several important discoveries of Richard Carrington. He can be credited with three first-rate discoveries—the solar flare, the differential rotation of the sun, the latitudinal drift of sunspots. Any one of these would have been sufficient for scientific immortality. But these outstanding discoveries could not save Carrington from an impending tragic fate. Stuart Clark has given a gripping account of Carrington's dramatic life in his book *The Sun Kings*. Here I sketch a brief summary of his life and point out its implications.

Figure 2.4 Gustav Spörer (1822–1895), who systematically studied the migration of sunspots from higher latitudes to lower latitudes with the sunspot cycle.

Carrington was the son of a successful brewer. As a young man, he set up his observing facilities at his own expense and carried out his astronomical work without holding any salaried position in any academic establishment. Apart from his own work that gave a definitive shape to the emerging new science of solar physics, he made sure that others making important contributions in the field were duly recognized. While on a trip to Germany, he went over to meet Schwabe, the discoverer of the sunspot cycle, and afterwards successfully campaigned that the highest honour of the Royal Astronomical Society—its Gold Medal—be bestowed on Schwabe. This happened in 1857. Two years later, Carrington himself was to receive this medal and was also elected to the Fellowship of the Royal Society at the young age of 34. As long as Carrington was carrying on his work without seeking any academic positions, his endeavours were fully appreciated and recognized. However, after his father's sudden death, Carrington found that the brewery was making too much demand on his time and he felt that he could continue his astronomical work only if he obtained an academic position. As soon as he began his efforts in securing such a position, suddenly troubles began.

Two successive openings were announced in Oxford University and Cambridge University, and Carrington applied for both. He was convinced that he was the most deserving person in England for both of these positions, but failed to get either. The selection for the Oxford position was particularly galling. It went to probably the least deserving of the applicants merely because he was the candidate of George Airy, the Astronomer Royal. In a letter to a more deserving applicant who did not get the job, Airy candidly admitted the shameless nepotism:

> Mr. Main's claims on me ... are like those of a son on the head of the family. This almost prevents me from saying a word in favour of any other person.[8]

Carrington was not the man to take these things lying down. He wrote incensed letters to several important persons in English academics, which did not endear him to others.

Afterwards things moved with the swiftness of a Greek tragedy. In a fit of rage, Carrington dismantled his telescope and all the laboratory equipments, which were then auctioned off. Soon after this, Carrington married a woman who was the mistress of another man, but had passed off that man as her brother. Having this extra complication to deal with in personal life certainly did not help matters. The end came when Carrington was barely 49. With a gap of only a few days, two deaths occurred under mysterious circumstances in the typical Agatha Christie setting of a quiet English countryside—the deaths of Carrington's wife and then Carrington himself. While things could not be established beyond doubt, circumstances hinted that the first death was probably a murder and the second death the suicide of the repentant murderer.

Carrington's short and tragic life exemplifies the famous Chinese curse: 'May you live in interesting times!' Carrington certainly did live in an interesting time—a time when science was changing its character rapidly. Up to the middle of the nineteenth century, some of the most important contributions in all branches of science were made by 'amateurs'—men who did not hold professorships in universities or salaried positions in research laboratories. Some earned their livelihood through other jobs and carried out their scientific work in evenings and weekends. Others were wealthy gentlemen who did not require a job. Probably the second half of the nineteenth century was the time when cutting-edge science decidedly started going beyond the reach of the amateur scientist. There were several reasons for this. Firstly, scientific equipments started becoming more elaborate and expensive. Secondly, as different branches of science became more and more technical, a prolonged period of training and apprenticeship became a prerequisite for a successful scientific career. Perhaps, most importantly, as the number of professional scientists increased and the competition became tougher, it became impossible for part-timers to win the race against full-timers.

Astronomy had a particularly glorious tradition of amateurs making important discoveries. Schwabe was an amateur whose income came from his pharmacy and who carried on his solar observations in his spare time. Carrington stands in a peculiar position in this tradition. He is the man who, more than anybody else, transformed the study of the sun into the modern technical discipline of solar physics that went beyond the reach of amateurs. After Carrington's time, I am not aware of any major discovery in solar physics made by an amateur. Amongst important solar astronomers at the end of the nineteenth century, Gustav Spörer began as a mathematics teacher but eventually obtained a job in the Potsdam Observatory, whereas Edward Maunder held a position in the Greenwich Observatory. Carrington must have seen these changes coming and wanted to reorganize his life in response to them. But he became a victim of the changes which ironically his own research heralded and paid dearly with his life. With one foot firmly planted in an era when important discoveries were being made by amateurs and the other foot planted in the era when science increasingly became a professional activity, the romantic and tragic figure of Richard Carrington marked a transition straddling the twin worlds of amateurs and professionals. With his demise, the age of amateurs in solar research ended forever.

2.4 The Solar Magnets

I have already mentioned in Section 1.4 that the crucial clue for understanding the sunspot cycle came when Hale discovered in 1908 that sunspots have strong magnetic fields. Most of us have enjoyed playing with magnets as children and have learnt some basic things about magnets in high school. Since magnetic fields are going to be very central in our story, let me recapitulate some basic things about them so that we all feel completely at home with magnetic fields.

The magnets with which you might have played as a child quite likely had the shape of a bar or a horseshoe. We shall see that magnets with both these shapes will appear in our story as we progress. A bar magnet has two opposite poles at its two ends. If you suspend the bar magnet, the north pole will point towards the north and the south pole towards the south. By playing around with bar magnets, you can quickly discover two very basic experimental laws of magnetism. (1) Like poles repel each other, whereas unlike poles attract. (2) You cannot separate single poles. If you break the bar magnetic to separate its two poles, you find that new poles appear in the regions where it has been broken so that each piece again has two poles. Let me begin by showing you how we can explain this by using the concept of magnetic field lines.

Although you cannot isolate a magnetic north pole, you can make a very long and thin magnetic needle—much longer than the size of your bar magnet. When you bring the north pole of this needle near your bar magnet, the south pole of the needle will be very far away and the force exerted by the bar magnet on the magnetic needle will basically be the force exerted on a north pole. The magnetic field can be technically defined as the force exerted on a unit north pole (let us skip the question of how the unit of a magnetic pole is defined). Since this force is a vector pointing in a particular direction, it can be indicated by an arrow having a direction. All the arrows near the north pole of the bar magnet should be in the outward direction because the north pole of the narrow needle will be repelled by this pole. On the other hand, the arrows near the south pole of the bar magnet should be in the inward direction to indicate attraction. If you draw such arrows all over (not only near the poles of the bar magnet), then you can keep joining the head of an arrow with the tail of the arrow in front of it. What you get in this way are called magnetic field lines. Figure 2.5(a) shows the magnetic field lines around a bar magnet. The concept of magnetic field lines is a tremendously useful concept introduced by Michael Faraday, and we shall be using this concept throughout the theoretical chapters of this book. Although the concept of magnetic field lines may seem like an abstract concept, you can make the magnetic field lines shown in Figure 2.5(a) visible through a simple trick. Put the bar magnet on a board and sprinkle some iron filings around it. If you now tap the board gently, you will find that the iron filings will take up positions to

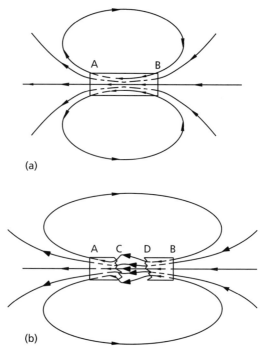

(a)

(b)

Figure 2.5 Magnetic field lines around (solid curves) and inside (broken curves) a bar magnet. (a) The bar magnet is a single piece. (b) The bar magnet is broken and the two pieces are moved slightly apart.

indicate the magnetic field lines. You may notice in Figure 2.5(a) that the magnetic field lines spread out when we move away from the poles of the bar magnet. In other words, these lines are more densely packed in regions where the magnetic field is stronger. The density of field lines in a region thus gives an indication of the strength of the magnetic field there, apart from the fact that the field lines indicate the direction of the magnetic field.

I already mentioned in Section 1.1 that Maxwell succeeded in putting all the basic laws of electromagnetism in the form of a set of elegant equations. Those of you who know vector calculus may look at a discussion of these equations in Appendix D. One of these equations is just the mathematical expression of the statement that magnetic field lines can never begin or end. In other words, a magnetic field line has to be continuous. In simple situations, a magnetic field line closes on itself, like a snake eating its own tail, though this need not always be the case in more complicated situations. We shall now see how this principle of continuity of magnetic field lines explains why new poles appear in the region of fissure when you break a bar magnet, or why isolated magnetic poles do not exist in nature. The continuity of magnetic field lines means that they have to continue inside the bar magnet, as shown in Figure 2.5(a). Now suppose we break the bar magnet AB into two pieces AC and DB, moving these pieces slightly apart. Figure 2.5(b) shows how the field lines would look now. Since the arrows are going inward into C, we conclude that C should now behave like a south pole. On the other hand, D becomes a north pole, with the arrows going outward from it. If we move the two pieces AC and DB far apart, they will simply become independent bar magnets, each having two poles. If we could somehow isolate a north pole, Figure 2.6 shows what the field lines would have looked like in that hypothetical situation. There is now no way of making the field lines continuous and the law of continuity of field lines will have to be violated. If we take the law of continuity of magnetic field lines as a fundamental law of nature (as encapsulated in one of Maxwell's equations), then we are forced to the conclusion that isolated magnetic poles are forbidden by this law.

Let us now come to the magnetic fields of sunspots discovered by Hale. Do we have both magnetic poles in sunspots? What do the magnetic field lines look like? Hale himself must have been bothered by

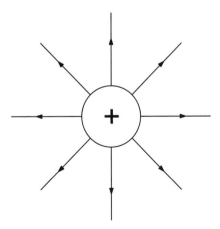

Figure 2.6 The magnetic field lines of a hypothetical isolated north magnetic pole.

these questions and, about a decade after his original discovery of magnetic fields in sunspots, he published a comprehensive long study of such questions in 1919 along with three collaborators—Nicholson, Ellerman and Joy.[9] Since the terms north pole and south pole can be very confusing when we apply them to the sun, let us instead use the terms positive and negative poles. Usually sunspots appear in pairs. Figure 1.3 shows a pair of sunspots. Hale and his collaborators discovered that, in such a sunspot pair, one spot is always a positive magnetic pole and the other a negative pole. Just as two opposite poles have to be present in a bar magnet, two opposite poles have to be present in sunspots as well—only they occur in two neighbouring sunspots and not usually within one sunspot.

Now we present something in Figure 2.7 which is absolutely fascinating. This figure is what is called a magnetogram map of the sun's disc at an instant of time. Such maps are produced by an instrument called magnetogram kept at the back of a solar telescope. The magnetogram measures the magnetic field at every point on the sun's surface. If the magnetic field has positive polarity at a point, then we put white colour at that point. On the other hand, black colour at a point indicates that the magnetic field has negative polarity there. The grey colour in a region means that the magnetogram failed to detect any magnetic field in that region. You can see that much of the solar surface has grey colour, which means that the magnetogram did not find any magnetic fields over most parts of the solar surface. The magnetic field is concentrated within the white and black patches in the magnetogram map. These patches are nothing but sunspots. If one could take a photograph of the sun at exactly the same time when the magnetogram map was made, you would find that the photograph showed sunspots in exactly the same regions where the magnetogram map shows white and black patches. A sunspot pair appears as a white patch and a black patch side by side in the magnetogram map, because one sunspot in the pair has positive polarity and the other negative polarity.

Before I discuss the significance of Figure 2.7 further, let me say a few words about conventions used in maps of the sun's disc. In maps of the earth's surface, we use the convention of taking north upwards. How do we define north in the earth's context? It is defined with respect to the earth's rotation axis, which pierces the earth's surface at two points which we call the north and south poles. Although the sun undergoes differential rotation, we can define its rotation axis uniquely.

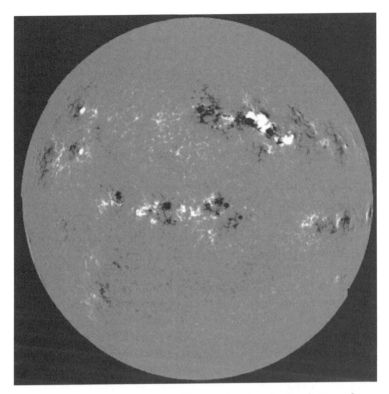

Figure 2.7 A magnetogram map of the sun showing the distribution of magnetic field over the sun's surface. See the text for explanation.

The rotation axis of the sun is approximately perpendicular to the plane in which the earth orbits around the sun. So, when we look at the sun from the earth, the sun's rotation axis appears approximately perpendicular to our line of sight. Hence, at any time, we see one hemisphere of the sun with the two poles at the two opposite edges. We normally present maps of the sun oriented in such a way that one pole is upwards and the other pole downwards. This is the case in Figure 2.7. The sun's equator would be a line going from left to right through the middle of the map.

Now look at Figure 2.7 more carefully. In the northern hemisphere (i.e. in the upper half of the map), you find that white patches appear on the right side of the black patches. This means that a sunspot pair in the northern hemisphere has the positive-polarity sunspot on the right and the negative-polarity sunspot on the left. The reverse of this holds in the southern hemisphere, where the positive-polarity sunspot is on the left and the negative-polarity sunspot on the right. The configurations of sunspot pairs remain this way during one cycle of 11 years. In the next cycle, however, the polarities of sunspot pairs get reversed. If we were to look at a magnetogram map made 11 years before or 11 years after the magnetogram map shown in Figure 2.7, we would find that the configurations of white and black patches would be reversed. In the northern hemisphere, we would find black patches on the right and white patches on the left. The southern hemisphere would have white patches on the right and black patches on the left. If we only consider the number of sunspots, we may think that the sunspot cycle has a period of 11 years. Solar astronomers refer to this 11-year cycle as the *Schwabe cycle* after its discoverer. On the other hand, we have to wait for two cycles spanning 22 years for the sun's magnetic field to come back to the same configuration. This magnetic cycle of period 22 years is called the *Hale cycle*. The observation that the sunspot pairs have opposite magnetic configurations in the two hemispheres and that these configurations change from one 11-year cycle to the next is known as *Hale's polarity law*.

While discussing Hale's epoch-making discoveries, I should mention that George Ellery Hale was probably the greatest observatory-builder in the history of astronomy. The precocious son of a wealthy and adoring father, the young Hale got his first telescope in the backyard of their palatial family home in Chicago. This beginning may be reminiscent of the beginning of Carrington's career. However, unlike Carrington who failed to get an academic appointment, Hale was lucky that a new university—the University of Chicago—came up almost next door to his family home and offered him a position. Hale decided that this new university needed the world's best observatory and persuaded a Chicago businessman, Charles Yerkes, to donate funds to build this observatory. In those days, astronomers believed that they should build observatories close to their home turf. Accordingly, the Yerkes Observatory was built at a distance of about 100 miles from the University of Chicago campus. Although this observatory boasted of state-of-the-art facilities including the great Yerkes telescope, which has remained to this day the world's largest refracting telescope (i.e. a telescope based on a lens rather than a mirror), Hale quickly realized that the countryside around Chicago is not particularly good for astronomical observations. I myself discovered it first-hand many years later when I was a graduate student at the University of Chicago and was supposed to do a summer project at Yerkes Observatory. My job was to study an eclipsing binary star during the two months of summer. An eclipsing binary consists of two stars going around each other. Its intensity varies with time because it becomes dimmer when the less bright star eclipses the brighter one. I was supposed to monitor how the intensity of this eclipsing binary varied during the two summer months. As it happened, there were barely four or five nights during the summer when the eclipsing binary could be seen clearly away from the clouds in the sky and I could take data. Thus ended my short career as observational astronomer.

Hale started arguing that observatories should be built in the best sites for astronomical observation rather than close to one's university. One may have the inconvenience of travelling there to make observations, but the scientific returns would be much higher. This is a notion which we now take for granted, but it was a radical suggestion in Hale's time. At a time when Los Angeles had not yet become the big city that it is today, Hale selected a hilltop known as Mount Wilson next to Los Angeles as the site for his observatory. With funding from the Carnegie Institution, he built the Mount Wilson Observatory which, more than any other observatory in the world, was responsible for a total transformation of astronomy in the first half of the twentieth century. After establishing this observatory, Hale had to hire astronomers and happened to handpick two unknown young men—Harlow Shapley and Edwin Hubble—who were later destined to change our perception of the universe in which we live. Hale also managed to obtain a donation from John D. Hooker to build the main telescope of the observatory—the 100-inch Hooker telescope—which was much more powerful than any other telescope that had existed earlier and which remained the world's largest telescope for several decades. It was this telescope which was used by Hubble to show that galaxies are the building blocks of the universe and that the universe is expanding with time. On Mount Wilson, Hale also set up his tower telescope for studying the sun, with which he carried out his fundamental investigations of magnetic fields in sunspots. Later in the book we shall discuss other important discoveries about the sun made from the Mount Wilson Observatory.

Hale also realized that the newly emerging science of astrophysics needed a journal and established *The Astrophysical Journal*. This journal

reached the pinnacle of glory during the long editorial stint of one of
Hale's successors at the University of Chicago, Subrahmanyan Chandrasekhar. It remains one of the world's main journals in this field and
many of the scientific discoveries discussed in this book were reported
in its pages. Unfortunately the end of Hale's life was tragic—though not
as tragic as the end of Carrington's life. Hale had a mental breakdown
when he was barely in his forties and started having hallucinations of a
'little elf' who would tell him how to conduct his life. His psychological
problems forced Hale to resign from the Directorship of Mount Wilson
Observatory at the age of 51, shortly after the publication of his great
1919 paper that established Hale's polarity law. He spent the last two
twilight decades of his life in seclusion as a patient of schizophrenia.

2.5 The Central Dogma

I now ask you to look once more at the magnetogram map in Figure 2.7.
You see many sunspot pairs seen as white and black patches next to each
other. Now you try to imagine straight lines joining the centres of two
sunspots in each sunspot pair. You can easily see that these straight lines
will be nearly parallel to the sun's equator. However, a very careful look
should convince you that these straight lines will not be strictly parallel to the equator. The right sunspot in a pair usually appears a little
closer to the equator than the left sunspot. As the sun rotates about
its rotation axis, any point on the surface of the sun moves from left
to right. The right sunspot in a pair, which is closer to the equator, is
in the forward direction with respect to the rotating sun and is called
the leading sunspot. The left sunspot in the pair is called the following
sunspot. What we are saying here is that the straight line joining the
centres of leading and following sunspots in a pair is usually not completely parallel to the equator, but is tilted at a small angle. It turns out
that this tilt angle is usually larger in sunspot pairs at higher latitudes. I
should point out that this is only an average law valid over many sunspots. Sometimes a particular sunspot pair at a low latitude may have
a large tilt or a particular sunspot pair at a high latitude may have a
small tilt. Even though we see a limited number of sunspot pairs in
Figure 2.7, it is still apparent that sunspot pairs at higher latitudes have
larger tilts. It was Alfred Joy, one of Hale's collaborators in the classic
investigation of sunspot magnetic fields at Mount Wilson Observatory,
who analysed the tilt angles of sunspot pairs to conclude that sunspot

pairs have small tilts with respect to the equatorial direction (the leading sunspots appearing closer to the equator) and that the tilt angle
increases with latitude. This result is known as *Joy's law*. I would request
readers not to forget this law as they proceed with the book. We shall
see that this law plays a tremendously important role in solar dynamo
theory.

We now come to a very crucial question: what do the magnetic field
lines look like in order to produce the magnetic configuration seen in
Figure 2.7? The answer to this question clearly requires some inspired
theoretical guesswork, because we know only of the magnetic field in
the sunspots and do not know anything about the nature of the magnetic field above or below the sun's surface. First of all, let us, for the
time being, forget about Joy's law, since it introduces extra complication. We will postpone until Chapter 5 a discussion of how the extra
complication of Joy's law arises. If both members of a sunspot pair are
always assumed to be exactly at the same latitude, what would the magnetic field lines be like? I now show in Figure 2.8(a) a theoretical guess
of how the magnetic field lines may look, in order to produce a typical
sunspot pair in the northern hemisphere. Basically, we have some magnetic field lines going around the sun's rotation axis in the solar interior.
Parts of these magnetic field lines are coming out of the sun's surface,
piercing the surface at two regions. It is clear that one of these regions

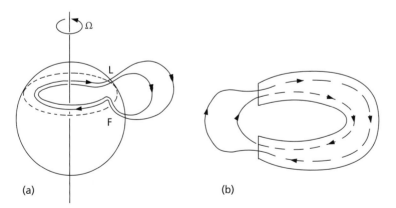

Figure 2.8 (a) A theoretical guess of how the magnetic field lines above and
below a sunspot pair (*L* and *F*) are expected to look like. (b) The magnetic field
lines of a horseshoe magnet.

has field lines coming out, which means that this region has positive magnetic polarity. The other region has negative magnetic polarity. If these two regions are the two sunspots, then we get two sunspots of opposite polarity side by side. In Figure 2.8(a), the leading sunspot indicated by *L* is shown to have positive polarity and the following sunspot indicated by *F* has negative polarity—exactly like what we see in the northern hemisphere in the magnetogram map Figure 2.7. While the magnetic field lines sketched in Figure 2.8(a) may explain the occurrence of sunspot pairs, we are now confronted with a deeper question: how do the magnetic field lines shown in Figure 2.8(a) arise? Before answering this question, we need to discuss some basics of MHD (abbreviation of magnetohydrodynamics, as pointed out in Section 1.4), which will be presented in Chapter 4. So you will have to wait until Chapter 5 for a complete answer to this question.

Figure 2.8(b) shows magnetic field lines of a horseshoe magnet. You can clearly see that the magnetic field lines above the sunspot pair seen in Figure 2.8(a) exactly resemble magnetic field lines inside the horseshoe magnet. In other words, we are reaching the conclusion that we should have something like a horseshoe magnet above a sunspot pair. The two sunspots are the two edges of this horseshoe magnet. Since we have the hot corona above the sun's surface, we expect the horseshoe magnet to be embedded in the corona. Within the last few decades, solar astronomers have discovered that the solar corona is really full of such horseshoe magnets. They are called *coronal magnetic loops*. Solar astronomers have also devised methods of 'photographing' these horseshoe magnets in the corona and have obtained some really spectacular images. Figure 2.9 shows some coronal magnetic loops. We shall discuss these loops more in Chapter 8, where I shall explain how their images are produced.

We now show in Figure 2.10(a) some magnetic field lines in the sun's interior which would give rise to the magnetic configuration seen in Figure 2.7. Parts of these magnetic field lines have to rise up and then come out of the sun's surface to produce the sunspot pairs. We have sketched the field lines as they would look before parts of them rise up and come out of the sun's surface. The field lines have to be in opposite directions in the two hemispheres, in order to explain the opposite polarities of sunspot pairs in the two hemispheres. In the technical jargon of dynamo theory, the magnetic field configuration shown in Figure 2.10(a) is called a *toroidal* magnetic field. On the other hand,

Figure 2.9 Coronal magnetic loops above the sun's surface. This image was taken in the extreme ultraviolet from the space mission Transition Region and Coronal Explorer (TRACE) on 6 November 1999. Credit: Stanford-Lockheed Institute for Space Research and NASA.

Figure 2.10(b) shows magnetic field lines which look somewhat like the field lines of a bar magnet embedded inside a sphere. We believe that the earth's magnetic field lines look somewhat like these. The magnetic configuration shown in Figure 2.10(b) is called a *poloidal* magnetic field. While learning about vectors, a student has to learn that two vectors can be added to give one resultant vector. In case the sun has a poloidal magnetic field in addition to the toroidal magnetic field, the sun's resultant magnetic field will be a combination of both. However, while discussing the solar dynamo problem, it is often useful to think of the toroidal and the poloidal components separately.[10]

In his seminal paper on dynamo theory in 1955, Parker suggested that the sunspot cycle is produced by an oscillation between the poloidal and the toroidal magnetic fields. We are familiar with many oscillatory phenomena around us. Think of a swinging pendulum. When the

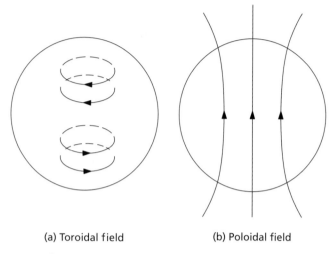

(a) Toroidal field   (b) Poloidal field

Figure 2.10 Sketches of (a) toroidal magnetic field lines; and (b) poloidal magnetic field lines.

pendulum is in the middle position, its potential energy is zero but its kinetic energy is maximum because it is moving fastest in that position. On the other hand, when the pendulum is in its extreme position, its kinetic energy is zero and the potential energy is maximum. In other words, there is an oscillation between potential and kinetic energies. Parker envisaged an oscillation between the poloidal and the toroidal magnetic fields arising out of some processes in which the poloidal field gives rise to the toroidal field and then the toroidal field, in turn, gives rise to the poloidal field by some other process. I should point out that the analogy between the pendulum and the solar dynamo should not be pushed too far because there are some profound differences. We do not have to supply energy to a pendulum for it to swing. We shall see later that the solar dynamo needs a continuous supply of energy in order to go on.

Parker's theoretical idea of an oscillation between the poloidal and the toroidal magnetic fields of the sun was truly extraordinary because very little was known about the sun's poloidal field at that time. In fact, in the same year 1955 in which Parker published his dynamo theory paper, a father–son pair of astronomers—Harold Babcock and Horace Babcock—discovered for the first time that there are weak opposite magnetic fields near the sun's two poles.[11] This established for the first time that the sun also does have a poloidal magnetic field like the earth, in addition to the strong magnetic fields inside sunspots. While the magnetic fields inside sunspots are typically as strong as 0.3 tesla, the magnetic fields near the poles were found to be weaker than 10^{-3} tesla. We have systematic data of the sun's polar magnetic fields only since the 1970s. So, only within the last few years, have we been in a position to address the question whether there is really an oscillation between the toroidal and the poloidal magnetic fields of the sun, as suggested by Parker. Figure 2.11 shows the relevant data. At the bottom of this figure, you see how the sunspot number has varied with time. Above the sunspot number plot, you see two curves showing how the sun's magnetic fields in the two poles varied during the same time. The data of the sun's polar fields come from the Wilcox Solar Observatory of Stanford University. Note that the magnetic fields of the sun's two poles have got to have opposite signs, as seen in Figure 2.11. When one polar field is positive, the other polar field is negative. The sun's polar fields are also seen to have a cycle of approximately 11 years, reversing from one cycle to the next. Around the years 1986 and 1997, the sunspot number was close to zero. Since sunspots form from the toroid field, these must be the times when the toroidal field was very weak. The upper two curves show that the polar fields were strongest precisely around these times. On the other hand, the polar fields were close to zero around 1989 and 2000 when the sunspot number reached its peak, implying a

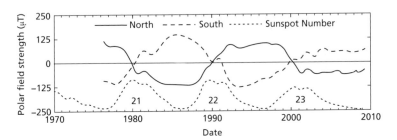

Figure 2.11 The sun's polar fields plotted along with the sunspot number starting from the mid-1970s. The two curves in the upper portion show the values of magnetic fields at the two poles (which have opposite signs) as measured at the Wilcox Solar Observatory. The lower dotted curve indicates the sunspot number. Credit: David Hathaway.

strong toroidal field. We thus see an oscillation between the poloidal and toroidal magnetic fields as Parker envisaged many decades earlier. It is simply mind-boggling to think how somebody could come up with such an idea years before it was established experimentally. Such a feat of theoretical intuition is the mark of a scientific genius of the first rank.

I am now in a position to state what I would call the central dogma of solar dynamo theory.

> The sun's magnetic field is made up of a toroidal magnetic field part and a poloidal magnetic field part. Sunspots form out of the toroidal magnetic field, whereas the polar magnetic field of the sun is a manifestation of the poloidal magnetic field. The sunspot cycle is produced by an oscillation between these two kinds of magnetic fields. The poloidal magnetic field gives rise to the toroidal magnetic field by some process. In its turn, the toroidal magnetic field also gives rise to the poloidal magnetic field by some other process, so that the cycle goes on.

If you wish to take away any short message from this book, this central dogma should be that message.

I would be very happy if my readers now want to know the details of the mechanisms by which the poloidal magnetic field gives rise to the toroidal magnetic field and then the toroidal magnetic field gives rise to the poloidal magnetic field. The next several chapters of the book will explain these mechanisms. If I could explain these mechanisms here with a few words, there would be no need of writing a whole book on this subject! These mechanisms are based on the principles of MHD and I shall be in a position to discuss these mechanisms only after I have given an introduction to MHD in Chapter 4.

When I started working in this field about three decades ago, there were a handful of solar physicists who steadfastly refused to accept that the sunspot cycle is due to a self-excited fluid dynamo. However, they failed to come up with a viable alternative to dynamo theory for explaining the sunspot cycle. Over the years, I have seen these critics of solar dynamo theory retire one after another and fade out from the scene. Their dissenting voices are no longer heard in the meeting rooms of important solar physics conferences. I am not aware of any respected solar physicist active in research today who doubts that solar dynamo theory holds the key to our understanding of the sunspot cycle.

While there appears to be a complete consensus about the central dogma amongst solar physicists at the present time, there is still no such consensus when we go to the details beyond the central dogma. For example, there are some disagreements and uncertainties about the detailed nature of the mechanism by which the toroidal magnetic field gives rise to the poloidal magnetic field. We shall discuss some of these disagreements in appropriate places. After all, such disagreements are always expected in the frontiers of a research field. If everybody agreed with everybody else all the time, then doing scientific research would have been much less fun!

2.6 A Road Map

Gene Parker often likes to make the statement: 'The sun is cleverer than all of us.' We see too many baffling phenomena on the sun which no theoretical physicist could dream of, had they not been observed. I have introduced some of the key phenomena connected with the sunspot cycle in the first two chapters of this book. Science begins with experiments. Astronomy is a peculiar science in which we cannot experiment directly with the objects we are interested in studying. So observations take the place of experiments. By now you have been exposed to the crucial observations with which any discussion of the sunspot cycle should begin.

In a mature branch of science, we usually have a core of theoretical concepts with which we explain various experimental results. We now come to the question of theoretically explaining different aspects of the sunspot cycle. This is certainly not an easy job. The rest of this book is meant to provide explanations for many of the phenomena which have been introduced in the first two chapters. However, we cannot immediately start with the theory of sunspot cycles because you need some theoretical background before we come to that. The next two chapters give this background. Since the sunspot cycle is produced by MHD processes in the sun's interior, you need to know something about the sun's interior and about MHD before we get into the theory of sunspot cycles. Chapter 3 will introduce you to the sun's interior and tell you how we can make many important inferences about the sun's interior on the basis of theoretical principles. Then I shall explain the basics of MHD in Chapter 4.

Chapters 5–7 constitute the central backbone of this book. It is in these chapters that I shall tell you why we have something like the sunspot cycle. A part of the central dogma is the question of how the toroidal magnetic field is generated from the poloidal magnetic field. We shall start addressing this issue towards the end of Chapter 4 and then Chapter 5 will explain how sunspots arise out of the toroidal magnetic field. After that we shall come to the remaining part of the central dogma—how the poloidal magnetic field is generated from the toroidal magnetic field and how the cyclic behaviour of the sun's magnetic field arises. Chapters 6 and 7 will be devoted to this crucial question lying at the heart of dynamo theory.

After we understand how the sunspot cycle is caused, we shall come to the question of why we have solar explosions in tandem with the sunspot cycle and how these explosions affect the earth—contradicting Lord Kelvin's calculations. This is the subject of Chapter 8. Finally, Chapter 9 will address the question why the sunspot cycle is not completely cyclic—why some cycles are stronger than the others and if we can predict the strength of a cycle before its onset.

Since I have already mentioned that disagreements exist about some details of solar dynamo theory, perhaps I should point out here that the topics which are controversial at the present time are more or less restricted to Chapter 7 and Chapter 9. I shall explain the nature of the disagreements in the appropriate places and give some idea of the alternative viewpoints. The rest of the book is pretty much standard stuff generally accepted by our scientific community at the present time.

3

Here Comes the Sun

3.1 Twinkle, Twinkle, Little Star

Why do stars shine? This has been one of the central questions in modern astrophysics and at least three scientists who made vital contributions towards answering this question were awarded Nobel prizes (Hans Bethe in 1967, William Fowler in 1983, Raymond Davis in 2002). But surprisingly the question why stars shine was rarely asked till about 200 years ago. Most scientists at the beginning of the nineteenth century thought that it was merely in the nature of stars to give out light and no further explanation was needed.

The beginning of the nineteenth century was still an age of incredible scientific beliefs. Nobody bothered about the origin of life because it was believed that living creatures could spontaneously arise out of inanimate matter. Serious scientists believed that bacteria were spontaneously generated in rotten meat broth, until Pasteur showed that this is not possible. Nobody was surprised that the sun and stars kept on pouring out unbelievable amounts of energy continuously, because nobody realized that it is not possible to create energy out of nothing.

It was in the 1840s that the principle of conservation of energy got established. It became clear that energy can be transformed from one form to another, but cannot be created or destroyed. Several scientists such as Julius Mayer, James Joule and Hermann Helmholtz proposed the energy conservation principle independently. Historians of science often puzzle over the question of simultaneous discoveries in science. Sometimes certain scientific ideas are almost 'in the air' and occur to several scientists independently. The discovery of the principle of conservation of energy is considered to be one of the famous examples of simultaneous discovery.

Once the energy conservation principle became established, it was an inescapable conclusion that some other form of energy in the interior of the sun must be getting converted into heat and light.

Hermann Helmholtz, one of the founders of the principle of conservation of energy, was the first scientist to consider this issue seriously, along with Lord Kelvin, a stalwart in the newly emerging science of thermodynamics. When we burn coal or wood, we get heat from chemical energy. Simple estimates convince one that the chemical energy of any combustible substance would be inadequate to account for the huge amount of heat the sun is producing over a long time. Kelvin and Helmholtz identified gravitational potential energy as the ultimate source of energy in the sun and suggested that it is this energy which is getting converted into other forms like heat and light.

We know that some gravitational potential energy is lost whenever an object falls in a gravitational field. Suppose you drop a brick from a height. The brick would have some initial potential energy by virtue of its position. As it falls, the potential energy keeps on decreasing, by being converted into the kinetic energy of falling motion. While the brick falls, the sum of the potential energy and the kinetic energy remains constant and is called the total mechanical energy. Everybody learns this well-known result, that the total mechanical energy of a falling object remains conserved, in high school science class and we assume that all readers know this. Eventually, as the brick hits the ground, the mechanical energy gets converted into other forms like heat and sound. The net result is that the brick's potential energy is transformed into other forms.

Kelvin and Helmholtz suggested that the sun and the other stars are slowly contracting in size due to their own gravity. As a result, the material inside stars must be falling towards the stellar centres, like the brick falling in the earth's gravitational field. This process would certainly cause a decrease in gravitational potential energy which, according to Kelvin and Helmholtz, is getting converted into heat and light to keep the stars shining. When one writes down mathematical equations based on the ideas of Kelvin and Helmholtz, one is led to an extremely simple, elegant and beautiful mathematical theory. The mathematical calculations suggest that the sun's radius has to decrease barely by about 30 metres in a year to produce all the heat and light that the sun is giving out.

At first sight, this may seem like a wonderful solution to the problem. If the sun's radius were decreasing at this very slow rate, even present-day instruments would find it very hard to detect this decrease. In the days of Kelvin and Helmholtz, there was certainly no way of determining whether the sun was shrinking at this rate or not. However, we get into trouble if we want the sun to shine for a very long time. Archbishop James Ussher used the chronology of *The Old Testament* to establish that God created the world on 23 October 4004 BC. If the world was really only 6000 years old, then there would have been no problem with the theory of Kelvin and Helmholtz. The sun's radius merely had to decrease by about 200 kilometres since its creation. However, geologists like Charles Lyell and evolutionary biologists like Charles Darwin were already arguing in the middle of the nineteenth century that the earth must be much, much older. After the discovery of radioactivity, scientists were finally able to date some of the oldest rocks accurately. There are rocks more than a billion years old. While 30 metres per year may seem a very slow rate of contraction to us, if the sun had to shine for a billion years, its radius would have to decrease by 3×10^{10} metres, which happens to be about 40 times the sun's actual radius! Clearly the Kelvin–Helmholtz theory could not be the correct theory for the sun's heat and light.

The final solution of the puzzle had to wait for the modern theory of the atom, which arose in the opening decades of the twentieth century. In 1897 J.J. Thomson discovered the tiny electron which must be a negatively charged constituent of the atom and showed that an atom is not an indivisible entity as many chemists had assumed. Finally Ernest Rutherford showed in 1911 that an atom consists of a compact positively charged nucleus with electrons orbiting around it. The atomic nucleus is made up of positively charged protons and electrically neutral neutrons. For example, the lightest element hydrogen has only a proton in its nucleus, whereas the next lightest element helium has a nucleus with two protons and two neutrons. When the mass of the helium nucleus was determined very carefully, it was found to be less than the combined mass of two protons and two neutrons. So, if one can put together two protons and two neutrons to create a helium nucleus, a small amount of mass has to leave the system. What happens to this mass? Even a person who knows no other equation of physics would have heard of Einstein's famous $E = mc^2$ equation, according to which mass can be converted into energy. The multiplication by the factor c^2 (where c is the speed of light) implies that even a small mass can give rise to a huge amount of energy. If we can build up a helium nucleus by bringing together two protons and two neutrons, then we can get quite a lot of energy from the small mass deficit.

In the 1920s several physicists began to argue that the energy of the sun must be due to nuclear reactions taking place in its interior. Einstein's famous equation tells us that about 5 billion kilograms of matter has to be converted into energy per second to account for the observed brightness of the sun. This may seem to you an enormously large amount of matter. But remember that the sun has a huge mass. The present age of the sun is estimated to be 4.5 billion years. During this time, the sun would have lost only 0.04 per cent of its mass due to conversion into energy in order to shine so brightly! So nuclear energy seems to be the best candidate for powering the sun and the stars. The question confronting astrophysicists in the 1920s was to find out the details of the nuclear reactions that must be taking place inside the sun and the other stars to convert appropriate amounts of matter into energy.

In the early decades of the twentieth century, astrophysicists realized that hydrogen is the most abundant element in the universe. The sun and most of the ordinary stars are primarily made up of hydrogen. If nuclear reactions could convert hydrogen into helium, then a small fraction of mass would be converted into energy and this could power the sun and the stars. As a first step in this process, two hydrogen nuclei (which are nothing but protons) have to come sufficiently close to each other so that a nuclear reaction takes place between them. The problem here is that protons or hydrogen nuclei have positive electric charges and repel each other. Only if one of them approaches the other with a very high relative speed, might they be able to come close enough overcoming the electrical repulsion and a nuclear reaction could take place. When a gas is heated to higher and higher temperatures, its constituent particles move around at higher and higher speeds. The question now is whether the hydrogen nuclei (or protons) in the hot core of the sun would be moving with sufficiently high speeds to overcome their electrical repulsion so that nuclear reactions could take place. First estimates suggested that even the centre of the sun is not hot enough for this to happen and many physicists doubted that nuclear reactions can take place at all in the sun.

An event that shook the foundations of physics in the middle of the 1920s was the formulation of quantum mechanics. One of the first things taught in college-level quantum mechanics is that a particle can penetrate through a barrier of repulsive force. This is called *quantum mechanical tunnelling*. In 1928 George Gamow, one of the greatest science popularizers of all time and the creator of Mr Tompkins, studied quantum mechanical tunnelling through the electrical repulsion of the atomic nuclei. When this tunnelling effect is taken into account, the sun's central temperature of 15 million degrees is found to be just enough for nuclear reactions to take place. As physicists kept learning more about nuclear reactions, around 1938 Hans Bethe was at last able to figure out the chain of nuclear reactions through which hydrogen must be converted to helium inside stars.

Apart from nuclear reactions, nobody has been able to come up with any other theoretical idea that can explain how the sun has kept producing huge amounts of energy for billions of years. However, suggesting that nuclear reactions produce the sun's energy on the basis of such considerations is clearly a negative argument. Since nothing else works, nuclear reactions have to produce the sun's energy! One important question is whether we can provide any compelling positive evidence that nuclear reactions are really taking place inside the sun. We shall come to this question a little bit later in this chapter.

Before leaving this topic of energy production in stars, I shall make one last comment about the Kelvin–Helmholtz theory. Kelvin and Helmholtz did not even know that the atom has a nucleus and certainly had no idea that the atomic nucleus can be a source of energy. Apart from not including this additional source of energy unknown at that time, Kelvin and Helmholtz made no mistakes in their theory. So, in a situation where nuclear reactions are not taking place, the Kelvin–Helmholtz theory should hold. We believe that new stars form due to contraction of gas clouds which exist in interstellar space. Initially nuclear reactions do not take place in the contracting clouds so that heat and light production takes place exactly in accordance with the Kelvin–Helmholtz theory. Eventually the central portion of the contracting gas cloud becomes sufficiently hot to start nuclear reactions and the gas cloud becomes a star which does not contract any further.

3.2 The Structure of Stars

One of the pillars of modern astrophysics is the theory of stellar structure. What is it? We expect that the temperature, density and pressure in the interior of a star should have their maximum values at the centre. Away from the centre, their values should be less than this maximum. In stellar structure theory, we solve a set of mathematical equations to calculate the values of temperature, density and pressure at different distances from the centre of a star. We definitely need a detailed knowledge of the energy generation rate by nuclear reactions in order to build a detailed model of a star. However, even during the late nineteenth century when scientists did not yet know the correct mechanism for energy generation inside stars, a few enquiring minds applied themselves to the problem of figuring out the set of equations which need to be solved for building a model of a star. Homer Lane, Jacob Emden and Karl Schwarzschild were some of the pioneers in this field of investigation. However, it was Sir Arthur Eddington who laid down the full foundations of the subject in the 1920s and wrote in 1926 a classic work *The Internal Constitution of Stars*, which had a tremendous influence on the development of stellar structure theory.

Some of us have the mistaken notion that it is easier to understand nearby objects compared to faraway objects. If that were the case, it would have been bad news for astrophysics, because astrophysicists always study faraway objects. It turns out that we understand an object better if its characteristics are governed by simpler physical laws. One of the reasons why astrophysicists had remarkable success in modelling stars is that a star is actually a rather simple system. It may come as a surprise to you that the structure of the sun is understood much better than the structure of the earth, even though the earth is underneath our feet. The materials inside the earth are in solid and liquid states, which are much more difficult to analyse than a gas. It was the great insight of Eddington to realize that the matter inside stars should behave like gases and should obey very simple physical laws (such as Boyle's and Charles's gas laws which you learn in high school). This is not at all obvious because the density in the central region of the sun is many times higher than that of water and most substances in terrestrial laboratories do not remain gaseous at such densities. Eddington argued correctly that the high temperature in the interior of a star would make the atoms move so fast that inter-atomic forces would be unable to bind them together to form solids or liquids. It is this insight of Eddington which showed that stars are simple objects to study and led to a rapid development of stellar structure theory. In the next chapter, we shall discuss that matter inside a star exists in the form of a special kind of gas: a plasma. The basic point is that stellar matter, to a very good approximation, obeys the ideal gas law (a combination of Boyle's and Charles's laws). For those readers who would like to know a little bit

Figure 3.1 Arthur Stanley Eddington (1882–1944), a pioneer in the study of stellar structure.

of the technical details of some important topics, I have given a few appendices at the end of the book. The ideal gas law is discussed in Appendix A.

One of the first concerns of stellar structure theory is to understand why most stars do not appear to be shrinking under their own gravity. After all, every chunk of material in a star is pulled towards the centre of the star by gravitational attraction. There must be something to balance this attraction. If the pressure in the interior region of the star is higher than the pressure in the outer region, then this excess pressure in the interior can balance gravity. The mathematical equation for this force balance is called the *hydrostatic equilibrium equation*. A simple form of this equation is taught in college physics courses. A slightly more complicated version of this is one of the basic mathematical equations of stellar structure. This equation is discussed in Appendix B.

In a nutshell, we need excess pressure in the interior of a star to balance its gravity. But what causes this excess pressure? According to basic gas laws, a hotter gas has a higher pressure. Nuclear reactions make the interior of a star hot and it is this high temperature in the interior of the star that is responsible for the excess pressure. We thus see that we need a heat source in the central region of the star in order to balance the star's gravity.

Eddington carried out his investigations at a time when details of nuclear reactions in the interior of a star were not known and the exact energy generation formula was not available. However, merely by assuming that there is some unknown source of heat in the interior of a star, Eddington was able to draw several far-reaching conclusions. For example, it was known to astronomers for some time that more massive stars are brighter. Eddington's theory gave an elegant explanation for that. A more massive star has a stronger gravitational field and a larger pressure excess in the central region is needed to balance it. Larger excess pressure can be produced if the central source of heat is stronger and this will definitely make the star brighter.

We now need to discuss one other aspect of stellar structure theory. The energy produced by nuclear reactions keeps coming out of the star in order to maintain a steady state. Since the nuclear reactions take place mainly in the central region of the star, the energy produced there has to be transported outward to the star's surface, from where the energy goes out in the form of heat and light. We now need to discuss how energy gets transferred from the central region of the star to its outer surface. High school physics textbooks usually list three modes of heat transfer: conduction, convection and radiation. It turns out that conduction is not important for heat transfer inside stars like the sun. So we have to focus our attention on convection and radiation.

Convection can take place only under some special conditions. You easily realize this if you heat a pan of water from below by putting it on a burner. Initially you see that there are no movements in the water and no convection. Only when the temperature difference between the bottom and the top of the water becomes sufficiently large, does convection start with the lighter hot water from the bottom coming up and the heavier cold water from the top going down. In other words, convection takes place in a fluid only when the temperature difference between the bottom and the top is high enough. If you do not mind using slightly technical language, we can say that the *temperature gradient* in a fluid has to be sufficiently large for convection to take place. Convection is an example of what is called a *hydrodynamic instability*. If the temperature gradient is small, then convection does not take place. Only on increasing the temperature gradient beyond a certain critical value, can we get convection in a fluid.

Figure 3.2 Karl Schwarzschild (1873–1916), who discovered the condition for convection inside stars.

In 1906 Karl Schwarzschild, working in Göttingen at that time, carried out a famous mathematical analysis of a gas in a gravitational field and showed how high the temperature gradient has to be in order to produce convection.[1] The fact that the temperature gradient of the gas has to be larger than the critical value given by Schwarzschild's analysis in order to start convection is known as the *Schwarzschild condition* for convection. The mathematical expression for this condition is given in Appendix B. We expect convection to take place in some particular region of a star if the temperature gradient there is larger than the critical value given by the Schwarzschild condition. On the other hand, if the temperature gradient is less than the critical value, then convection will not take place. The only mechanism left for heat transfer in such a region of a star is radiation. We now have to say a few words about how radiation causes heat transfer.

The transfer of heat and light from the sun's surface to the earth's surface is a particularly simple example of heat transfer by radiation. We know that radiation consists of a stream of photons. Since the space between the sun's surface and the earth's surface is virtually devoid of any matter that can absorb radiation, photons race through this space like bullets moving at speed c (the same speed of light which appears in the equation $E = mc^2$) and take about eight minutes to reach the earth after leaving the sun's surface.

The transfer of heat from the very hot centre of the sun to its much cooler surface by radiation is a much more complicated process, because there are atoms around which can absorb radiation. The energy generated by nuclear reactions comes out in the form of radiation (or photons). These photons are absorbed by the surrounding atoms. The atoms which absorbed radiation get excited and will emit photons. These photons are, in turn, again absorbed by atoms and this process goes on. The net result is that the energy generated at the centre of the sun diffuses outward through the repeated absorptions and re-emissions of photons by surrounding atoms. This process is called *radiative transfer*, which is certainly much less efficient and much slower than the transfer of heat through empty space by radiation. You may not at once realize how slow this process is. Astrophysicists have estimated that some energy generated at the centre of the sun would take about 30,000 years to reach the surface by this process of radiative transfer. The sunlight that you see out of your window was produced by nuclear reactions that took place in the centre of the sun at a time when our ancestors were fighting with woolly mammoths and sabre-toothed tigers! If nuclear reactions in the centre of the sun were to stop suddenly for some reason, we would probably become aware of that many thousands of years later.

How slow or fast the radiative transfer is depends on how much resistance the surrounding matter offers to the radiation passing through it. If the matter is nearly transparent, then the radiation would pass more easily. On the other hand, if the surrounding matter is fairly opaque, then radiative transfer becomes much less efficient. The technical word to indicate the opaqueness of matter is *opacity*. In order to build a model of a star, we need to study how energy is transported outward from the centre of the star by radiative transfer and this requires a knowledge of the opacity of the surrounding matter. Hence the opacity of matter inside the star is an important ingredient for building models of stars.

The opacity of stellar matter is calculated theoretically. Let me just give you a rough idea of how this is done. Barely a year after Rutherford proposed the nuclear model of the atom, Niels Bohr realized that electrons should occupy only a few discrete orbits inside the atom if we are to explain why atoms have spectral lines. In the Bohr model of the atom, an electron moves from a lower orbit to a higher orbit when a photon is absorbed by the atom. On the other hand, an electron in a higher orbit can jump to a lower orbit by emitting a photon. Quantum mechanics, developed a few years later, is more or less in conformity with this Bohr picture of absorption and emission of photons by atoms. To calculate the opacity of stellar matter, we have to do a rather complicated bit of book-keeping: figuring out which atoms are there, how many electrons are in which orbits in these atoms and how likely these atoms are to absorb a photon by pushing up an electron from a lower orbit to a higher orbit. These are complicated calculations, but some groups have been doing these calculations methodically and diligently for many years. Other scientists normally use the values of opacity calculated by these groups in their work.

3.3 Peeping into the Interior of the Sun

By now we have discussed all the important bits of physics that go into making models of stars. I now need to recapitulate and give you an idea how the model calculation actually proceeds.

First of all, the matter inside the star has to obey the ideal gas law (the combination of Boyle's and Charles's laws) at least approximately. Then we need to know the nuclear energy generation rate and the opacity of this matter at various temperatures and densities. Many groups have calculated these things and these data are usually available in the form of tables which anyone can use. Once we have all these data, we need to solve the hydrostatic equilibrium equation and the heat transfer equation to build the model of the star. As I already pointed out, there is a little bit of complication in applying the heat transfer equation, because the heat transfer equations for radiative transfer and convection are quite different, and we do not know which equation to use until we have an idea whether heat is transferred by radiative transfer or by convection. Since convection is much more efficient than radiative transfer, we can usually forget about radiative transfer in regions where convection takes place. The usual procedure is the following. First we proceed by using the heat transfer equation by radiative transfer and keep calculating the temperature gradient at every step. If we find that the temperature gradient in a region is higher than what is needed for convection according to the Schwarzschild condition, then we know that convection has to be present there and we replace the heat transfer equation by radiative transfer with the heat transfer equation by convection. This is roughly how models of stellar structure are constructed.

Within the last 40 or 50 years, many astrophysicists have worked out the structure of the sun. With improvements in the input data (nuclear energy generation rate and opacity) and in modelling, results of different groups have converged and we now have what is called the *standard model* of the sun. According to this standard model, the temperature at the centre of the sun is 15 million degrees, whereas the density there is about 150 times the density of water. It is amazing to think that we now know what the values of density and temperature are at different points in the interior of the sun. But you may have a justifiable concern. This knowledge has come entirely from theoretical calculations. How can we be so absolutely sure that our theories are really correct? This is a very valid question. Physics is an experimental science. Sometimes even our most careful theoretical calculations give wrong results because we might have overlooked something. So it is always a good idea to check our theoretical calculations against experiment. Can we check the standard model of the sun by independent experiments? I shall discuss this point later in this chapter. Here let me just mention that it has indeed become possible in the last few years to check the standard model of the sun through experiments. The result? The standard model of the sun has come out with flying colours and is now regarded as a triumph of theoretical astrophysics.

Let us now come to the very important question whether heat is transferred by radiative transfer or convection in the interior of the sun. We get an answer to this question when calculating the structure of the sun. The sun has a radius of about 700,000 kilometres. The energy generated in the central region of the sun by nuclear reactions is transported outward by radiative transfer up to a radial distance of about 500,000 kilometres. This is called the *radiative core* of the sun. Beyond the radiative core, heat is transported by convection to the solar surface. This region, which has a thickness of about 200,000 kilometres is called the *convection zone* of the sun. Figure 3.3 shows a sketch of what the sun's interior is like, indicating the locations of the radiative core

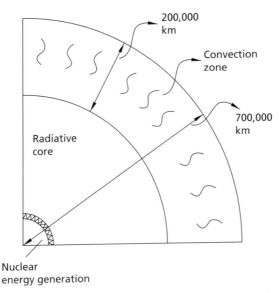

200,000
km

Convection
zone

700,000
km

Radiative
core

Nuclear
energy generation

Figure 3.3 The interior of the sun according to the standard model based on the stellar structure equations.

and the convection zone. You may wonder why convection takes place in the outer layers of the sun. Let me give you a rough answer. As I explained above, the opacity arises due to atoms absorbing photons. I have also mentioned that matter at high temperatures tends to become ionized and electrons come out of the atom. If an atom is completely ionized, that is, if no electrons are left in the atom, then we would have a bare nucleus and the absorption of photons is no longer possible. Basically we would have more opacity when more electrons are left inside the atoms. Then photons can be absorbed by these electrons moving to higher orbits. Since there will be more electrons left inside atoms at lower temperatures, we expect the opacity to go up at lower temperatures. There is a law known as *Kramers's law of opacity* which gives the mathematical relation between temperature and opacity. Since the temperature in the outer region of the sun is less than the temperature in the central region, the opacity and resistance to radiative transfer is more in the outer region. As a result, the temperature gradient would have been very high there if energy were transferred by radiative transfer. Such a high temperature gradient tends to cross the critical value for starting convection. That is why we have the convection zone in the outer part of the sun.

One of the important conclusions of the standard model of the sun is that heat is transferred by convection in the layers just below the sun's surface. You can get visual evidence of this just by looking at the sun's surface, which looks like the top of a boiling liquid when photographed by a powerful telescope. Take a careful look at Figure 1.3 showing a pair of sunspots, in which you can also see the sun's surface around the sunspots. The bright granular patches seen in the photograph are called *granules*. They are nothing but regions where hot gas is coming up. The typical size of a granule is about 700 kilometres. The granules are separated by dark lanes, which are regions where cold gas is going down. Instead of a static photograph, if you were to look at a movie of the solar surface, you would find the granular pattern changing completely in a few minutes. Old granules would fade out and new granules would appear.

A look at a photograph like Figure 1.3 convinces us that sunspots are basically patches of magnetic field concentration sitting in a region of the sun where vigorous convection is going on. As we proceed, we shall see that an understanding of the interaction between convection and magnetic fields is going to be really crucial for our story. In fact, we shall find that the whole of the dynamo process for generating the 11-year sunspot cycle takes place within the convection zone. Particularly, the bottom of the convection zone will turn out to be the place where many things that are important for the dynamo process happen.

3.4 How Will the Sun Die?

Now we take a very brief look at a question which may not be directly relevant to the main theme of this book. But this question may arise in the minds of some readers. Attempts to answer this question led to some of the most dramatic developments in the history of astrophysics over the last century.

We have discussed that the inward gravitational attraction of the sun is balanced by the excess pressure in the interior arising out of the heat generated by nuclear reactions. We know that hydrogen is getting converted into helium due to these nuclear reactions. As a result, the store of hydrogen in the sun's core is depleted continuously and a time will surely come when there will be no hydrogen left in the sun's core for any more nuclear reactions. What will happen then? From a practical viewpoint, there is no immediate urgency in answering this question. The sun's present age is about 4.5 billion years and astrophysicists have estimated that the sun has enough hydrogen to go for at least 5 billion years more. But what lies at the end of the road?

Sirius, the brightest star in the sky, has a faint mysterious companion star known as Sirius B. When astronomers studied Sirius B carefully at the beginning of the twentieth century, they arrived at the startling conclusion that its size was comparable to the size of the earth, although its mass was close to the mass of the sun. This implied that the density of matter inside the star was more than a million times the density of water. Such stars of small size and enormous density are called *white dwarfs*. Sirius B was the second white dwarf to be discovered and astronomers soon found many other white dwarfs.

When a normal star runs out of material for nuclear reactions (often called 'nuclear fuel') in its core, it can no longer produce the heat that would provide the excess pressure to balance gravity. Hence it is expected that the star will start shrinking due to its own gravity. Could a white dwarf be a possible final state of a star which has exhausted its nuclear fuel and has shrunk due to its own gravity? Detailed research has indeed supported this viewpoint. After about 5 billion years, our sun will become a white dwarf. I should point out that there is a small twist to this story. When hydrogen is exhausted in the core of the sun, at first the sun will bloat up and enter a temporary phase known as the red giant phase, for reasons which we need not discuss here. Once the red giant phase is over, the sun will start its final journey towards becoming a white dwarf. A white dwarf is basically a dead star in which no more nuclear reactions are taking place. The original star shrinks quite a bit due to its own gravity to become a white dwarf, but after that it does not shrink further. How does the inward gravitational attraction eventually get balanced inside the white dwarf to stop further shrinking? This question could only be answered after the advent of quantum mechanics in the mid-1920s.

According to the ideal gas law (the combination of Boyle's and Charles's laws, discussed in Appendix A), the pressure of a gas becomes zero when its absolute temperature (in kelvin) falls to zero. We know that the pressure arises due to random motions of gas particles. At the absolute zero of temperature, the gas particles stop moving altogether and there cannot be any pressure. This is the result from classical physics. So, if there is no source of energy to maintain non-zero temperature, the gas in the core of a star will have zero pressure and will not be able to counteract gravity. Quantum mechanics modifies this result in a very important way. One of the famous results of quantum mechanics is Heisenberg's uncertainty principle. It is a very profound principle and we cannot get into a full discussion of it here. One of the consequences of this principle is the following. If we know the position of a particle fairly precisely, then it has to have a non-zero velocity. Hence, according to this principle, particles will have velocities due to quantum uncertainty even when the temperature is zero. As a result, there will be a pressure even at zero absolute temperature, especially if the density is high and particles do not have much room around them to make a large uncertainty in position possible. One other principle of quantum mechanics that helps us is Pauli's exclusion principle. According to this principle, two electrons are forbidden to have exactly the same position and velocity. When matter has very high density and many electrons are forced to occupy very proximate positions, this exclusion principle dictates that some electrons must have rather high velocities. The pressure arising out of such quantum mechanical considerations is called *degeneracy pressure*. One hopes that this degeneracy pressure may be able to balance gravity in white dwarfs even when there are no nuclear reactions and no source of heat.

Ralph Fowler, a colleague of Eddington at Cambridge University, was the first person to look into this question and showed in 1926 how degeneracy pressure can balance gravity inside a white dwarf. But a big surprise was just around the corner. In 1930 Subrahmanyan Chandrasekhar, a brilliant Indian student, had won a scholarship to study at Cambridge and was on his way to England from India by ship. His plan was to do his PhD thesis under Fowler's guidance. During the long sea voyage, Chandrasekhar carefully went through Fowler's mathematical derivations. Chandrasekhar knew that an electron can never move at a speed greater than the speed of light c according to Einstein's special theory of relativity. While deriving the degeneracy pressure, Fowler had clearly not taken this into account. Electrons were free to move at any large speed in Fowler's theory and could produce the pressure to balance the white dwarf's gravity. When Chandrasekhar included the restriction that electrons should never move at speeds larger than c, he found that the pressure he got was less than what Fowler was getting by allowing the electrons to move at any speed. Could this reduced

Figure 3.4 Subrahmanyan Chandrasekhar (1910–1995), whose work made us understand how stars die.

pressure also balance the gravity inside the white dwarf? The calculations of the 20-year-old Chandrasekhar yielded a totally unexpected result, which shook the foundations of astrophysics and took several decades to be accepted generally in the astrophysical community.[2]

Chandrasekhar found that his revised degeneracy pressure could balance the gravity inside a white dwarf only if the mass of the white dwarf was less than a limiting value. This limiting value of mass, now known as the *Chandrasekhar mass limit*, was found to be about 1.4 times the mass of the sun. So the sun will have no difficulty in finding final peace as a white dwarf. But astronomers knew that many stars have masses several times the mass of the sun. What will happen to these stars when they exhaust their nuclear fuel? If degeneracy pressure could not balance their gravity, then the amazing possibility that stared astronomers in the face was that such stars will go on shrinking to become point objects and the gravitational field will be so strong that light or any other signal will not be able to escape from the surrounding region. We now call such an object a black hole. Most astronomers in the 1930s, however, were not ready to accept such a possibility.

Eddington, with whom Chandrasekhar had become personally fairly close after his arrival in Cambridge, was one person who felt that such a conclusion was completely absurd.[3] Eddington admitted that he could not find any mistakes in Chandrasekhar's calculations: 'If one takes the mathematical derivation of the relativistic degeneracy formula as given in astronomical papers, no fault is to be found.' So Eddington concluded: 'I think there should be a law of Nature to prevent a star from behaving in this absurd way!' In a meeting of the Royal Astronomical Society in 1934 that has since become the stuff of astrophysical folklore, there was a clash of the titans. After Chandrasekhar presented his work on the mass limit of white dwarfs, Eddington rose to argue that Chandrasekhar's theory had to be wrong at some fundamental level because it led to such an apparently absurd conclusion. It was, however, an uneven battle. Eddington was at that time at the peak of his fame as the world's most distinguished astrophysicist and many astrophysicists took his statements to be the last word on any question connected with the structure of stars. On the other hand, Chandrasekhar was a very young man and the astrophysicists present in that fateful meeting had no way of knowing that in a few years he would come to be regarded as the greatest theoretical astrophysicist the world has ever known. After this public battle, Chandrasekhar took a most extraordinary decision. He probably had an inner conviction that truth will finally prevail. In order not to prolong this unpleasant debate, Chandrasekhar decided not to work on the structure of stars any more. He wrote a famous monograph on the subject presenting his point of view and then moved on to work in other areas of astrophysics. Only several decades after this, did astronomers gradually realize that Chandrasekhar's work helped them to make sense of a vast amount of astronomical data that continued to accumulate. In 1983 Chandrasekhar was finally awarded the Nobel Prize in physics for 'theoretical studies of the structure and evolution of stars'—about half a century after he had stopped working in that field!

In fairness to Eddington, we have to admit that the idea of a black hole is not an easily palatable idea. Newton's theory of universal gravity is justly regarded as one of the greatest landmarks in the history of science. However, this theory cannot give an adequate account of the situation in which a star has shrunk to virtually zero size and light cannot escape its gravitational field. In 1915 Einstein developed his theory of general relativity, which provided a new and deeper formulation of gravity. In general relativity, gravity is regarded as curvature of four-dimensional spacetime (I shall refrain from getting into a detailed discussion here as to what this means). After a star has collapsed to a point, we need to find the nature of spacetime around it to understand the behaviour of the system. Karl Schwarzschild, who discovered the condition for convection inside stars, worked out a famous solution of Einstein's equation of general relativity giving the structure of spacetime around a collapsed star. Einstein's equation of general relativity is so complicated that Einstein himself thought at first that it would not be easy to work out a solution of this equation. Schwarzschild proved him wrong within a few months. The circumstances under which this extraordinary work was done were rather tragic. It was the time of World War I. Schwarzschild was a patriotic German and offered to serve in the army. One would have expected that a person of his age (about 40) and scientific reputation would not serve in the dangerous regions near the front. But Schwarzschild insisted on serving at the front. There he contracted a fatal disease. Schwarzschild read Einstein's newly published paper on general relativity while in his sickbed and worked out his solution when he was feeling slightly better. But soon afterwards the disease snuffed out his life at the age of 42. Judging by the amount of outstanding contributions Schwarzschild had made to astrophysics in his short life, one can only wonder how much more he would have achieved had he lived longer.

In the first few years after Einstein's work on general relativity was published, very few people understood the theory or its significance. Had Eddington been a person who did not know general relativity or did not know that this theory provided a systematic explanation of the nature of a collapsed star, one could excuse him for his unwillingness to accept Chandrasekhar's work. But the supreme irony is that Eddington was one of the early enthusiasts about general relativity who did much to champion this theory in England. He also led the famous first experiment to check the validity of the theory. One of the remarkable predictions of general relativity is that light should bend while passing through a gravitational field. A star near the edge of the sun will appear pushed further away from the sun due to this bending of light. We cannot, of course, normally see a star near the edge of the sun. But, during a total solar eclipse, stars around the sun may become visible. One has to take a photograph of these stars during an eclipse and compare this with a regular photograph of the stars to verify if some stars have indeed shifted away from the sun. A total solar eclipse in 1919, visible from an island off the coast of Africa, provided the right opportunity. Eddington led an expedition to study this eclipse and announced to the world that the stars shifted exactly by the amount predicted by general relativity. It is this discovery which flashed in the newspaper headlines around the world and made Einstein a household name.

Even with a full command over general relativity, Eddington could not accept the possibility of a black hole. The Eddington–Chandrasekhar debate seemed to have left lasting marks on the personalities of both men. I have already mentioned that Chandrasekhar left the study of stellar structure to work in other areas of astrophysics and eventually to scale greater heights. Eddington, on the other hand, after charting out the course of stellar research for so many years, became increasingly immersed in a twilight world of strange scientific fantasies. Already by the time of the Eddington–Chandrasekhar debate, Eddington had started working on what he called the 'fundamental theory', trying to unify some diverse areas of physics. While Eddington might have considered Chandrasekhar's work wrong, virtually the entire physics community regarded Eddington's fundamental theory as sheer nonsense. Bitterly disappointed at the negative reception of his fundamental theory, Eddington became increasingly isolated from the physics community and died a broken man—sometimes an object of ridicule by younger physicists for his fundamental theory.

Before I leave the topic of stellar death, I should mention that I have deliberately left out one important point in order keep our narrative direct and simple. Around the time when the Eddington-Chandrasekhar debate took place, some astronomers realized that there can be another possible end stage of a stellar collapse besides the white dwarf: what is called a neutron star. A star having mass greater than the limiting mass of a white dwarf can become a neutron star. Here I shall not discuss neutron stars any further, except to mention that they also have an upper mass limit. If a dead star has a mass more than this mass limit, then the stellar collapse has to go all the way, leading to the formation of a black hole.

3.5 Neutrinos: Mysterious Messengers from the Sun's Core

After this brief detour to take a glimpse of the dying sun 5 billion years into the future, let us again come back to the present living sun. We have argued that the huge energy output of the sun must be produced by nuclear reactions in the sun's core because nobody could think up any viable alternative. We now come to the question of whether we can find an experimental proof that nuclear reactions are actually taking place in the sun's interior. In the last few decades, a messenger from the sun's core has finally provided the proof: the mysterious particle neutrino.

Let us begin at the beginning. Some radioactive nuclei decay by throwing out an electron. This kind of radioactive decay is called β-decay. The initial nucleus before the decay had a particular amount of energy. The final nucleus after the decay must also have a particular lower amount of energy. We expect the ejected electron to have an energy equal to the difference in energies between the initial and final nuclei. In other words, we would expect all electrons in a particular β-decay process to have the same energy. When physicists carried out careful experiments, they found that this was not the case. There was a maximum energy, with no electron found to have energy more than that. But the values of the electron energy were found to cover the entire range from zero to the maximum energy. The only possibility, which Wolfgang Pauli proposed in 1930, is that there must be another particle ejected with the electron and the electron is sharing its energy with this particle. This proposed particle was named neutrino and could only be detected more than a decade later. The main difficulty in detecting neutrinos is that they interact extremely weakly with matter. A beam of neutrinos would pass through thousands of kilometres of lead without being attenuated very much. To be able to detect a particle, the particle has to interact with a detector. If the particle passes through the detector without interacting with the detector at all, then we would not know that the particle had been there. Hence one needs fairly sensitive and sophisticated experiments to detect neutrinos.

The nuclear reactions in the core of the sun produce a huge number of neutrinos. Since the whole sun is virtually transparent to these neutrinos, they flow freely in all directions from the core of the sun. Believe it or not, 65 billion neutrinos pass through every square centimetre

Figure 3.5 The experimental setup for detecting neutrinos deep underground in the Homestake gold mine. Raymond Davis (1914–2006) won the Nobel Prize in 2002 for this experiment.

of the earth that faces the sun in every second! If one could capture a few of them and confirm their presence, then that would provide a proof that nuclear reactions are indeed taking place in the sun. In the 1960s Raymond Davis started the first experimental effort to detect solar neutrinos. The experiment was conducted in the Homestake gold mine, about one mile below the surface of the earth. The main reason behind conducting this experiment so deep underground was that cosmic rays and all other disturbances above the earth's surface were completely cut off, allowing just a pure beam of neutrinos to reach there. The detector was a huge tank of capacity 100,000 gallons filled with the cleaning fluid carbon tetrachloride. Very occasionally a neutrino was expected to react with a nucleus of chlorine to produce a nucleus of argon. Since it was a very rare reaction, the fluid tank had to be of enormous size in order to increase the likelihood of a reaction taking place. Once in a few weeks, Davis would analyse the fluid in the whole tank to count the number of argon atoms which had been produced and thereby to estimate the strength of the neutrino beam. This was certainly much more challenging than looking for the proverbial needle in a haystack.

After running the experiment for a few years, Davis and his colleagues announced the results.[4] They did detect neutrinos, but the strength of the neutrino beam was found to be about one-third of what was theoretically expected. This was both good news and bad news. The good news was the evidence that nuclear reactions were taking place inside the sun and were producing the neutrino beam. The bad news was that the one-third discrepancy implied that there must be something wrong with the theory or the experiment or both. This mismatch between theory and experiment was known as the *solar neutrino problem* and remained one of the most celebrated unsolved problems in physics for several decades. For more than two decades, the Davis experiment was the only experiment for detecting solar neutrinos. Subsequently several other experiments using entirely different detection methods started. But all of the experiments found neutrino beams to be somewhat weaker than what was theoretically predicted. There had to be something wrong somewhere.

Physicists had known for some time that there are three kinds of neutrinos. It was extremely difficult to determine their masses. The only thing that physicists could say was that their masses were either exactly zero (as in the case of photon, the particle of light) or much smaller than the mass of the electron, the other lightest known particle. Now, if the neutrinos have non-zero masses, one gets a very interesting theoretical result. One type of neutrino can be spontaneously converted to other types. All the neutrinos produced in the core of the sun are so-called electron neutrinos and all the solar neutrino experiments also detected only electron neutrinos. If the neutrinos had non-zero mass, then on the way from the sun to the earth the electron neutrinos would continuously get converted into other kinds of neutrinos and then re-converted back again into electron neutrinos. As a result of these neutrino oscillations, the final neutrino beam reaching the earth would consist of one-third electron neutrinos, one-third muon neutrinos and one-third tau neutrinos (muon neutrinos and tau neutrinos being the other two kinds of neutrinos). If that were the case and the detectors on earth detected only electron neutrinos, then we would naturally expect the experimentally determined beam strength to be one-third of the theoretically predicted beam strength.

Although this possible resolution of the solar neutrino problem was known for many years, most physicists suspected that neutrinos have exactly zero mass. Then one type of neutrino could not get converted into the other types and this explanation could not be the correct explanation. A decisive experiment in 2002 finally showed that neutrino oscillations do take place, implying that neutrinos must be having non-zero masses. This finally drew the curtain on the solar neutrino problem. From being one of the celebrated unsolved problems in physics, the solar neutrino problem became a celebrated solved problem. However, even before the solar neutrino problem was solved, the very existence of solar neutrinos convinced the astrophysics community that nuclear reactions are taking place in the sun and the energy of the sun indeed comes from nuclear reactions.

3.6 The Vibrating Sun

Every serious student taking up physics major in college has to study the wonderful *Feynman Lectures on Physics*. The text of the lectures delivered by Richard Feynman in Caltech was prepared by three authors: Feynman himself, Robert Leighton and Mathew Sands. The second author, Robert Leighton, was Feynman's close friend and colleague in Caltech for many years. Leighton was a very versatile physicist and made contributions in different areas of physics. But his most important works are in the study of the sun.

In 1962 Leighton was studying the surface of the sun along with two of his PhD students Robert Noyes and George Simon, using an instrument for accurately measuring any velocities present on the solar surface. There is a standard technique in astronomy to find out if an astronomical object is moving towards us or away from us. The technique is Doppler effect, which is routinely used by police patrols to find out if you are speeding. Basically, astronomers take a spectrum of the astronomical source and check if the spectral lines are shifted from their normal positions. Any avid reader of popular science would know that Hubble discovered the expansion of the universe when he found that spectral lines of most galaxies were shifted towards the red, implying that the galaxies are moving away from us. On the other hand, if an astronomical object is moving towards us, its spectral lines would be

Figure 3.6 Robert Leighton (1919–1997), who discovered that the sun is vibrating all the time.

shifted towards the blue. Leighton and his students fabricated a fairly sensitive instrument for detecting very small Doppler shifts in the spectra of the solar surface. They attached this instrument to the solar tower telescope of the Mount Wilson Observatory, from where Hale had discovered the magnetic field of sunspots about half a century earlier. Leighton, Noyes and Simon made the astounding discovery that a point on the sun's surface continuously goes up and down—just like a man's chest going up and down as he breathes in and out.[5] The typical time a point on the sun's surface would take to complete one cycle of up-and-down motions is about 5 minutes.

When a drum vibrates, it involves vibrations in several frequencies combined together. Only some devices like a tuning fork vibrate at a single frequency. Vibrations of most vibrating objects are made up of vibrations in many frequencies known as the normal frequencies of the objects. At the beginning of the nineteenth century, the French mathematician Joseph Fourier, while studying the problem of heat flow, had developed a mathematical technique for combining vibrations of different frequencies. When we are given a vibrational pattern (such as information about change in air pressure with time when a sound wave passes through air), we can apply the mathematical technique known as *Fourier analysis* to find out which frequencies are present in the vibrational pattern. About a decade after the original discovery of the vibrations at the solar surface, Franz-Ludwig Deubner succeeded in gathering enough data about these vibrations to make a Fourier analysis possible. He announced in 1975 that these vibrations on the sun's surface are made up of vibrations at certain definite frequencies—just as in the case of a vibrating drum.[6]

By now solar astronomers have carefully measured thousands of frequencies present in the vibrations of the sun's surface. What information do these frequencies give us? Let us consider a stretched string. When it vibrates, the normal frequencies present in the vibration are determined by the length of the string, its mass per unit length and the tension along the string. Given the values of the normal frequencies, a competent physicist should be able to infer various characteristics of the string. In exactly the same way, it is possible to find out various information about the interior of the sun from the values of the sun's frequencies.

The simplest way to proceed is as follows. We have the standard model of the sun giving the values of density, pressure and temperature at all interior points. From this standard model, one can theoretically calculate the normal frequencies of the sun and then compare these with the frequencies measured by solar astronomers. When this programme was carried out, the result was astounding. The agreement between the theoretically calculated normal frequencies and the frequencies measured by observers turned out to be so fantastically good that there remained little doubt that the standard model of the sun really did give the correct values of density, pressure and temperature in the sun's interior.

This experimental confirmation of the sun's standard model is undoubtedly one of the biggest triumphs of modern astrophysics. Stellar structure is an important part of modern astrophysics which every student of astrophysics has to learn. Astrophysicists have been constructing detailed models of stars for several decades by solving the stellar structure equations. Since this is a very complex subject requiring different kinds of inputs that go into building the models, there were always lingering doubts in the minds of astrophysicists as to how good these models are. The triumphant validation of the standard model of the sun gives us confidence that stellar structure theory is indeed capable of giving very good models of stars. The solar neutrinos confirmed that nuclear reactions are truly taking place in the core of the sun. The vibrations of the sun have now demonstrated that the density, pressure and temperature in the sun's interior points calculated from stellar structure equations are very close to their actual values.

The science of studying the sun's vibrations is called *helioseismology*. In ordinary seismology, we use seismic waves (the waves produced during earthquakes) to get information about the earth's interior. The sun's vibrations cause waves to go around the sun and it is these waves which give us different kinds of information about the sun's interior. We now turn to one of the most spectacular achievements of helioseismology which provides a major ingredient in building models of the solar dynamo.

3.7 Rotation Rate of the Sun's Interior

We have discussed in Section 2.1 how Galileo used the changing positions of sunspots with time to conclude that the sun rotates about its axis and then how Carrington discovered that the rotation rate of the sun varies from lower to higher latitudes. Like Galileo, Carrington also used sunspots as markers on the solar surface to make this tremendously important discovery. In the twentieth century, however, astronomers developed techniques for measuring the Doppler shifts of spectral lines near the edge of the sun. As we expect, one side of the sun was found to be moving away from us and the other side approaching us. Careful measurements showed how the rotation rate varies from the equator to the pole. Near the faster rotating equator, a point takes about 25 days to go around the rotation axis of the sun. On the other hand, a point near the pole may take nearly 35 days to go around the rotation axis. After astronomers carefully measured the rotation rate all over the sun's surface, they started to wonder how the interior of the sun rotates. For theoretical reasons which we shall not discuss here, some astronomers conjectured that the sun's core may be rotating much faster than its surface. However, if you had asked an astronomer around 1975 whether we would ever be able to measure the rotation rate of the sun's interior, you would probably get the answer

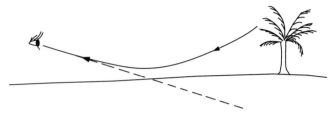

Figure 3.7 The path of a light ray above the hot desert sand, illustrating how a mirage is produced.

that this was never going to be possible. Then some new developments in helioseismology changed the situation all of a sudden.

To understand how this was possible, we consider a question which at first may seem completely unrelated to what we are discussing: how mirages form in deserts. When the sun's rays heat up the desert surface, the air just above the surface may become hotter than the air a little bit above. It turns out that the speed of light in hot air is more than that in cold air. When a ray of light moves from regions of colder air to regions of hotter air, it continuously keeps getting refracted. Figure 3.7 shows the path which such a ray of light starting from the top of a date palm tree would follow. It gets refracted up from the hotter air near the surface. When this upward moving ray reaches the eye of a person, he or she mistakenly thinks that the date palm tree is reflected in a body of water.

In Figure 3.8 we consider a point A on the surface of the vibrating sun which must be sending out waves due to these vibrations in different directions. I should point out that any point of the sun should act as a source of waves, but let us focus our attention on the point A. These waves are certainly not light waves, but turn out to be essentially the same as the acoustic waves which carry sound in the earth's atmosphere, except that the frequencies of these waves in the sun are usually much lower than the frequencies we need for sound to be audible to our ears. For these waves in the sun also, the speed increases as the waves travel deeper into the hotter regions of the sun. So these waves also can get refracted upwards when they try to move down, exactly like the ray of light from the top of the date palm tree. ABC is the path of such a wave travelling from A in the counter-clockwise direction. It turns out that such waves get reflected from the sun's surface. So the wave we are considering gets reflected from C and keeps tracing up repeated arcs like ABC. Only when such a wave going around the

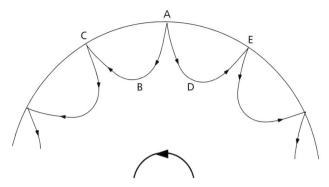

Figure 3.8 Counter-clockwise and clockwise moving waves starting from the point A on the sun's surface. See text for explanations.

sun interferes with the starting wave in an appropriately constructive way, do we get a normal vibration of the sun corresponding to a normal frequency. In Figure 3.8 we also show a clockwise propagating wave *ADE*, which is exactly similar to the wave *ABC* except that it is propagating in the opposite direction. If the counter-clockwise and the clockwise directions are completely symmetrical, then both the waves shown in Figure 3.8 should have the same normal frequency. But are these opposite directions completely symmetrical?

If the sun were not rotating, then there would be nothing to destroy the symmetry. Then the counter-clockwise and the clockwise directions would be completely symmetrical. The normal frequencies of waves propagating in both the directions would be the same. Now assume that the sun is rotating around an axis perpendicular to Figure 3.8. The thick arrow indicates the rotation direction of the sun. You can easily see that the counter-clockwise and the clockwise directions are now no longer symmetrical. The counter-clockwise wave propagates in the direction of the sun's rotation, whereas the clockwise wave propagates in the direction opposite to the sun's rotation. Since the two opposite directions are now no longer symmetrical, we should not expect the counter-clockwise and the clockwise waves to have the same normal frequency any more. In other words, the single normal frequency that we would have had in the absence of rotation gets split into two normal frequencies. The amount of splitting between the frequencies would depend on the rotation rate. When the rotation varies over the sun, the splitting would primarily depend on the rotation rate at the turning points *B* and *D*.

Douglas Gough and some of his co-workers were among the first people to realize that the rotation rate in the sun's interior may be measured this way. When they carefully analysed the helioseismology data, they indeed found that many normal frequencies had the kind of splitting that would be expected from rotation. The different normal frequencies correspond to waves that have their lower turning points (points like B or D in Figure 3.8) in different regions of sun. Hence, from the splittings of many different normal frequencies, one can figure out rotation rates in different regions in the interior of the sun.

The first tentative results on the rotation rate in the sun's interior came in the early 1980s when I was a PhD student in Chicago and had started working with Gene Parker. I still vividly remember that both Parker and I first heard about these new results from Bernard Durney of National Solar Observatory in Tucson, who was visiting Chicago for a couple of days. Some of Durney's colleagues in Tucson were involved in this work. Sitting in Parker's office, Durney described the new results to Parker and me. Parker had not heard about these new results before and wanted to know the methodology by which they had been obtained. Durney started explaining. But he did not have to say much. Parker had an almost uncanny capability of grasping new ideas of physics at lightning speed. I have rarely come across others who can match him. As soon as Durney started explaining the basics, Parker could immediately grasp how the rotation rates in different interior points of the sun could be determined and was terribly excited. In fact, during the four years I worked with Parker, I had rarely seen him getting so excited about any new result. However, I myself was completely at a loss. I did not understand the methodology and did not follow much of the scientific discussion between Durney and Parker. After we came out of Parker's office, I requested Durney to explain the methodology to me again. Durney did not brush off this request from a green PhD student who did not yet have any published research at that time. He came to my office and again explained everything very kindly and patiently. Although I did not follow all the things he said, I could sense that something tremendously important was happening. That is why I still remember many of our conversations of that day so vividly even after 30 years. About one sunspot cycle later—that is, about 11 years later—Durney and I would both be involved in developing a new type of dynamo model of the sunspot cycle in response to the discoveries in helioseismology. But we shall come to that later.

On the basis of helioseismology, solar astronomers produced maps showing how the rotation rate varied from point to point in the sun's interior. As higher quality data kept coming, these maps kept improving for about a decade until by the late 1990s they more or less converged to the form that is accepted today. Figure 3.9 shows a slice of the sun's interior made by a plane passing through the sun's axis. All the points in this plane which have the same rotation rate are connected by a curve. Several such curves are shown in Figure 3.9. The dashed line indicates the bottom of the sun's convection zone at a depth of 200,000 kilometres below the surface. You can see in Figure 3.9 that several curves of equal rotation rate lie very close to the each other at the bottom of the convection zone. This means that the rotation rate changes very quickly from point to point as we move upwards or downwards near the bottom of the convection zone. In more technical language,

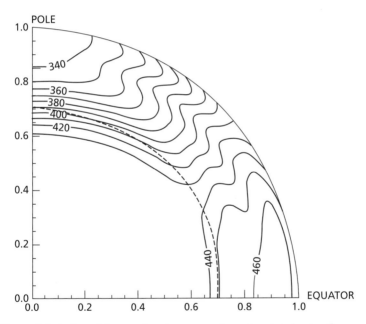

Figure 3.9 A slice of the sun showing curves through points having the same rotation rate. The dashed circle indicates the bottom of the convection zone. The rotation frequencies in nanohertz are given with the curves (1 nanohertz means 10^{-9} rotation per second). I may mention that frequencies of 340 and 450 nanohertz respectively correspond to rotation periods of 34.0 and 25.7 days.

we say that there is a strong concentration of differential rotation at the bottom of the convection zone. Edward Spiegel and Jean-Paul Zahn, two cultured astronomers who knew Greek and Latin (unlike most of their counterparts), coined the name *tachocline* for this layer of concentrated differential rotation at the bottom of the convection zone. We have already discussed that the bottom of the convection zone is a very interesting place where the mode of heat transfer changes from radiative transfer below to convection above. Now we find that this bottom is even more interesting as a region which has a concentration of strong differential rotation. We shall see later that the bottom of the convection zone plays a very important role in the dynamo process that produces the sunspot cycle.

You may want to know why the sun has the kind of complicated differential rotation that we see in Figure 3.9. I shall not attempt to answer that question in this book, because the theory of differential rotation is an immensely complicated subject. Let me only mention that very intricate theoretical calculations carried out by scientists such as Leonid Kitchatinov and Günther Rüdiger have given us some ideas how the differential rotation of the sun arises.

3.8　A Postscript

One might naively think that the interior of a star would be a region most inaccessible for scientific investigations. It is amazing to think that on the basis of the laws of physics discovered in our laboratories we could develop a fairly complete picture of what the interior of the sun is like and that this picture could be confirmed so triumphantly by developments such as the solar neutrino experiment and helioseismology.

A star is a very large object and you may think that we would not need any knowledge of physics at the atomic level to understand a star. As should be clear from the discussions in this chapter, it is the developments in quantum mechanics, atomic physics and nuclear physics which enabled us to calculate the nuclear energy generation rate and opacity, without which detailed models of stars would not have been possible. In astrophysics, we repeatedly see that we very often need a knowledge of the smallest things in order to understand the biggest things. It is this interplay between the very small and the very big which makes modern astrophysics such a fascinating scientific discipline.

4

The Fourth State of Matter

4.1 An Electromagnetic Interlude

After discussing in the previous chapter what the interior of the sun must be like, we would like to address the question of how the magnetic field arises inside the sun to produce the 11-year sunspot cycle. However, you need to know certain things about magnetic fields in a plasma before we can come to that question. The aim of this chapter is to introduce you to the unusual behaviour of magnetic fields in plasmas, while refreshing your memory about certain electromagnetic laws which you are likely to have studied in high school. To those readers who are completely at home with the basic electromagnetic laws, my apologies for the first few pages of this chapter which may not have anything new for such readers.

There is a story about a shepherd boy in ancient Greece who noticed that his iron-tipped staff was repeatedly being attracted to a particular stone. Whether this story about the discovery of magnetism is true or not, the existence of magnetism was known to some of the ancient civilizations like Greece and China. Many of us as children have been mystified and fascinated to watch how magnets attract small pieces of iron. In the Middle Ages, lodestones were believed to be associated with black magic and the occult. Sailors were terrified by stories of seaside mountains made of lodestone which could pull out all the iron nails in a ship, causing it to fall apart. In Section 1.4, I mentioned William Gilbert whose 1600 book *de Magnete*—one of the great classics in the history of science—did much to bring magnetism into the realm of science out of the realm of magic.

Apart from magnetism, the ancient Greeks are credited with another discovery. When amber was rubbed against fur, it was found that they attracted each other. This phenomenon—which is called electricity—was not thought to have any connection with magnets until about 200 years ago. It was around 1800 that Alessandro Volta was able to construct the first primitive version of a battery (known as Volta's pile) and showed that electricity can be made to flow through a metal wire in the form of an electric current. Once scientists knew how to produce electric currents, the connection between electricity and magnetism was just waiting to be discovered. The actual discovery came in 1820 when Hans Christian Oersted, while preparing a lecture for his students, noticed that a magnetic needle placed near a current-carrying wire got deflected. It was clear that electricity in motion (that is what an electric current is) can affect magnets. Soon after Oersted's discovery, André Marie Ampere figured out the exact mathematical equation which connects the electric current and the magnetic field produced by it. By applying Ampere's equation, one can calculate the magnetic field around a current-carrying wire.[1]

Often in the history of physics there comes a decisive moment when two phenomena, previously regarded as unconnected, are found to have deep connections. Oersted's discovery that moving electricity affects magnets led to the synthesis of electricity and magnetism into the single science of electromagnetism. Just as moving electricity produces magnetic effects, moving magnets can produce electrical effects by inducing electric currents in circuits. This phenomenon—which is called *electromagnetic induction* and which, as we shall see later, lies at the heart of our story of sunspots—was discovered in 1831 by Michael Faraday, one of greatest experimental geniuses of all time. Faraday carried out many experiments systematically to elucidate various aspects of electromagnetic phenomena. Since Faraday came from a very poor family and never had any formal college education, he never had an opportunity of learning mathematics properly. He formulated many of his revolutionary ideas in non-mathematical everyday language. His example shows that it is often possible to explain truly epoch-making new ideas in physics without using advanced mathematics.

The previous sentence is not meant to deny the important role of mathematics in physics. Sometimes mathematical formulations of physical phenomena make inter-relations amongst various phenomena clearer and give rise to completely new ideas. This happened when James Clerk Maxwell succeeded in combining all the available knowledge of electromagnetism into a set of beautiful and elegant equations. In 1866 Maxwell showed that these equations allow the existence of a kind of electromagnetic wave. Maxwell managed to calculate the speed of this wave, for which purpose he needed the values of some electrical and magnetic quantities that could be measured in the laboratory. When the speed of the wave was calculated using these values, it was found to be fantastically close to the speed of light known at that time. The conclusion was inescapable: light is a form of electromagnetic wave. This was the second great synthesis in this field. Earlier electricity and magnetism were synthesized into electromagnetism. Now electromagnetism and light got synthesized in one of the grandest edifices of classical physics.

While all these ground-breaking developments were taking place in electromagnetism, most physicists thought that only a few substances showed electrical or magnetic effects. Many substances around us—such as wood—seem to lack any electrical or magnetic properties. Physicists knew that gravity is universal in the sense that all substances produce gravitational attraction. Electromagnetism, in spite of the series of brilliant discoveries in the nineteenth century, appeared restricted to only certain substances and seemed not to have the universal character of gravity. This view changed only in 1911 with one of most landmark developments in the history of physics—Rutherford's establishment of the nuclear model of the atom. Physicists came to realize that all atoms have positively charged nuclei with negatively charged electrons going around them due to their electrical attraction. If the total negative charge of the electrons in an atom is equal to the positive charge of the nucleus, then the atom is electrically neutral. This is supposed to be the normal state of an atom. If all the atoms of a substance are electrically neutral, then the substance may appear not to have any visible electrical properties. However, the atoms inside the substance would not have been possible without the electrical attraction between their nuclei and the circling electrons. Physicists also discovered that most of the elementary particles such as electrons and protons are like tiny magnets themselves. Hence atoms and molecules made out of them often tend to have magnetic properties (though not always). A substance which appears non-magnetic simply has the atomic magnets randomly pointing in all different directions so that the substance as a whole shows no net magnetism.

The current view of physicists is that electromagnetism is a universal interaction. Substances which seem to lack electromagnetic properties merely have the charges and magnets at the atomic scale so fantastically balanced that we do not become aware of their existence. In fact, physicists now think that there are only four fundamental interactions in the universe—gravitational, electromagnetic, strong and weak. Unlike electromagnetism in which the effects of positive charge may be balanced by the effects of negative charge, gravity cannot be 'screened'. Although gravity is intrinsically much weaker than the other three interactions, in large systems like planets or stars which have equal amounts of positive and negative charges, it is gravity which turns out to be the dominant interaction. As it happens, two of the fundamental interactions—weak and strong—only act over very short distances and are usually confined within the atomic nucleus. Strong interaction is responsible for overpowering the electrical repulsion amongst the protons within the nucleus so that the nucleus does not fall apart. Radioactive decay of atomic nuclei is often caused by weak interaction. When we consider systems ranging in size between the size of atomic nuclei on the one hand and that of mountains on the other hand, it is usually the electromagnetic interaction which is the most important interaction for these intermediate-sized systems. Atoms, molecules and the solid objects we see around us all belong to this intermediate category.

Atoms and molecules form due to electromagnetic interactions. However, we know that the usual Newtonian mechanics that students learn in high school does not hold for atoms and molecules. We need to apply quantum mechanics. When electromagnetic interaction is treated with the help of quantum mechanics, it can sometimes camouflage itself in such a way that we may not recognize it as electromagnetic interaction. Why is the roof above my head holding on and not falling on my head? Why am I able to move my muscles in such a way as to raise my hand? It is the electromagnetic interaction which is ultimately behind all these, although we may not recognize this fact immediately.

It is a measure of the genius of nineteenth-century physicists that they discovered all the important principles of electromagnetism long before the universal nature of electromagnetism was realized. When twentieth-century physicists discovered that electromagnetism is so universal, they had all the principles and equations in their hands ready for application. With the establishment of Rutherford's model of the nuclear atom in 1911, another fact became clear. While the normal state of an atom may be one in which the total negative charge of the electrons is equal to the positive charge of the nucleus, an atom may sometimes lose a few of its electrons. We now consider this situation.

4.2 What is a Plasma?

We learn in elementary school that matter exists in three possible states—solid, liquid and gas. For common substances like water, we are familiar with all of the states. It is usually the temperature that determines the state in which we find a particular substance.

Count Rumford, while supervising the boring of cannons at a factory in Bavaria, noticed in 1798 that the movement of the boring machines gave rise to a seemingly endless quantity of heat. Careful investigations led him to the conclusion that heat is a form of motion. However, until atomic theory reached a state of maturity a few decades later, nobody was sure as to what kind of motion heat was. With the development of atomic theory, it became clear that all substances are made of atoms and molecules. Heat is essentially the motion of the atoms and molecules. As the temperature is raised, this motion becomes more vigorous. We also know that atoms and molecules can exert forces amongst themselves. As we pointed out, these forces ultimately originate from electromagnetic interaction, although this may not always be easily recognizable. Sometimes the inter-molecular forces bind many molecules together to form a solid. This is what happens in ice, which results from the inter-molecular forces between water molecules. On the other hand, heat makes molecules want to move around freely without becoming bound to each other.

Within a substance, the inter-molecular force and the thermal motion oppose each other. When the temperature is low, the inter-molecular force wins over and binds the molecules together to produce a solid. If heat is added and the temperature increases, thermal motion increases, and when it becomes comparable in importance to the inter-molecular force, the solid turns to liquid. Because of thermal motion, the molecules in a liquid are not bound in a rigid grid. However, the inter-molecular force ensures that the molecules remain close to each other in a packed state. On raising the temperature further, the thermal motion gets the upper hand and the molecules simply run around freely without the inter-molecular force being able to bind them together any more. This is the gaseous state.

A normal gas consists of freely moving molecules. When a gas is heated to higher and higher temperatures, the forces binding atoms into the molecules yield to the thermal motions and we are left with a gas made up of atoms rather than molecules. Then some of the

Figure 4.1 Meghnad Saha (1893–1956), whose equation explains how an ordinary gas gets converted into a plasma on increasing the temperature.

electrons start coming out of the atoms. This is the process known as *thermal ionization*. The positively charged remnant of an atom from which one or more electrons have come out is called the ion. Eventually, at very high temperatures, we have a collection of freely moving electrons and ions. Matter in this state is called a plasma. Since many properties of matter in the plasma state are sufficiently different from the properties of ordinary gases, the plasma state is often regarded as the fourth state of matter.

Different particles in a gas have different amounts of energy. A more energetic particle moves faster than a less energetic particle. There are methods of physics by which one can find out the number of particles in a gas which will have a certain amount of energy. The branch of physics in which such calculations are done is called statistical mechanics. When an atom in a gas acquires sufficient energy to knock off an electron by overcoming the electrical attraction of the nucleus, we get a positively charged ion. Meghnad Saha working in India applied the principles of statistical mechanics to calculate how often this would happen at a certain temperature. He was able to come up with the now famous equation—the *Saha equation*—which can be used to calculate the fraction of atoms in a gas at a certain temperature and pressure which will be broken up into ions and electrons.[2] With the increase of temperature, more and more atoms are ionized and eventually the gas is fully converted into the plasma state, with nearly all the atoms broken into ions and electrons. From the Saha equation and some other considerations, it seems that more than 99% of ordinary matter in the universe should be in the plasma state.[3]

Richard Feynman makes a striking remark at the beginning of his lectures on the electromagnetic field in the famous *Feynman Lectures on Physics*:

> If you were standing at arm's length from someone and each of you had *one percent* more electrons than protons, the repelling force would be incredible. How great? Enough to lift the Empire State Building? No! To lift Mount Everest? No! The repulsion would be enough to lift a 'weight' equal to the entire earth![4]

Let us consider a small volume element inside a plasma which is, however, large enough to have many positively charged ions and negatively charged electrons within it. Usually the total positive charge within this volume will be very nearly equal to the total negative charge. Feynman's example should convince you that even a small charge imbalance would give rise to enormous forces. For example, if there is a deficiency of negative charge in a region of a plasma, the excess positive charge there will quickly attract electrons from the surrounding region and the charge deficiency will get rectified pretty fast. In other words, a small volume element in a plasma will have no net charge even though it is made up of charged particles. Now, if a small volume element of a plasma has no net charge like a small volume element of an ordinary gas, do we conclude that the physical properties of the plasma are identical with the physical properties of the ordinary gas? A little reflection will show that this is not the case. Suppose we apply an electric field— by some means such as inserting two electrodes in the gas or the plasma and then connecting the two electrodes to the two ends of a battery. Since an ordinary gas (like air) is a bad conductor of electricity, nothing much will happen, unless the electric field is extremely strong to cause an electric discharge like lightning. On the other hand, an electric field applied to the plasma will immediately produce an effect, by making the ions move in the direction of the electric field and by making the electrons move in the opposite direction. In other words, an electric field readily induces a current in the plasma.

From the above considerations, we can conclude that plasmas are good conductors of electricity. If we consider a volume element in a plasma through which an electric current is flowing, we shall find that the amounts of positive and negative charges at an instant of time are the same so that the volume element will have no net charge. But oppositely charged particles move in opposite directions as shown in Figure 4.2, producing the current. Now remember Oersted's discovery that electric currents can deflect magnetic needles. In other words, electric currents produce magnetic fields around them. We would certainly expect currents in plasmas to give rise to magnetic fields. We indeed very often find magnetic fields in plasmas, suggesting that electric currents must be flowing through them. Stars and galaxies made up of plasma very often have magnetic fields. Otherwise, I would not be writing this book about the sun's magnetic field and sunspots. However, although the earth and the sun have magnetic fields existing over vast regions of space, we do not see similar electric fields existing over vast distances. The reason should be obvious now. Magnetic fields are produced by electric currents, whereas electric fields are produced by electric charges. Since it is quite common for plasmas to have electric currents, it is not surprising to find magnetic fields resulting from these currents. On the other hand, the lack of net electric charge within volumes of plasma ensures that there would be no electric fields. Since it is the electric field which drives a current in a plasma, you may wonder if there would be currents in the plasma at all if there are no electric

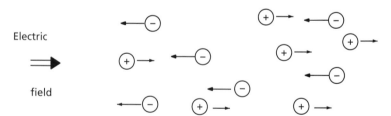

Figure 4.2 A diagram showing how a current can arise in a plasma due to an electric field, making positively charged ions and negatively charged electrons move in opposite directions.

fields. The point to remember is that a plasma is an extremely good conductor of electricity. So even a very weak electric field can cause a strong current in the plasma. It is no wonder that magnetic fields are usually much more important than electric fields in plasmas.

4.3 The Reality of Fields

Almost everybody in the ancient world held (and unfortunately, quite a few even today hold) the view that stars can influence our lives. From the time of the Renaissance, the idea of something exerting an influence on something else at a large distance started falling into general disfavour. However, against this general trend, Newton postulated his theory of universal gravitation. This proved to be a triumphant theory which achieved a kind of success in explaining planetary motions that no previous physical theory ever had. In spite of this success, Newton himself apparently used to feel rather uncomfortable about the idea of gravity acting between bodies separated by vast regions of space.

Newton's mechanics gave rise to a view that most natural phenomena can be understood in terms of bodies pushing and pulling each other. This can be called the mechanical view. Einstein wrote a brilliant popular book *The Evolution of Physics* with his younger collaborator Leopold Infeld. This book gives a marvellous account of the rise and decline of this mechanical view. When heat could be understood in terms of motions of atoms and was brought within the fold of the mechanical view, it was undoubtedly a great triumph for this point of view. An ambitious programme at the beginning of the nineteenth century was to reduce all of physics to a mechanical view. Einstein and Infeld described how this programme eventually floundered due to its failure to explain all of the electromagnetic phenomena from the mechanical view, and how a tremendously important new concept—the concept of field— came to dominate physics. Although physicists initially attempted to give mechanical interpretations of various properties of fields, eventually fields were taken as ultimate real entities endowed with certain properties which could not be explained in terms of anything else. Here I shall point out some issues that are relevant to our discussion. For a full discussion of this revolution in physics, readers should turn to *The Evolution of Physics*, one of the most profound popular science books ever written.

Henry Cavendish and Charles Coulomb in the eighteenth century were amongst the first scientists to make careful measurements of electrical repulsion between like charges and electrical attraction between unlike charges. The force law turned out to be the same inverse square law as in Newton's law of gravitation. In other words, the force at twice the distance should be a quarter in strength. It thus turned out that gravitation was not a unique kind of force. The nature of the electrical force is exactly similar. While we cannot separate the two poles of a magnet, it appears that forces between magnets can also be explained by assuming that each pole produces a magnetic force obeying the inverse square law.

Now, how do masses, electric charges or magnetic poles exert such forces around them? One approach is to say that we only concern ourselves with the force exerted by one mass on another mass, one charge on another charge or one magnetic pole on another magnetic pole. We know that these forces obey the inverse square law and we take it at its face value, without asking the more probing question of how these forces arise. This approach is called the action-at-a-distance approach. However, a mechanical view will require explanations of these forces in terms of pulls and pushes of something. Can we provide such mechanical explanations?

To answer this question, we turn to a very important concept introduced by Faraday: the concept of field lines or lines of force. I have already introduced this concept for magnetic fields in Section 2.4. In Figures 4.3(a) and 4.3(b) we see the electric field lines between two equally strong unlike and like charges. I shall point out one big difference between electric and magnetic field lines, which follows from the basic Maxwell's equations. While magnetic field lines cannot begin or end (Section 2.4), electric field lines begin from positive charges and end at negative charges. You can see that the electric field lines spread as we move away from the charges to regions where the electric fields are weaker. This suggests that the density of field lines is an indication of the strength of the field. In a region of strong field, we have more field lines next to each other. On the other hand, fewer field lines near each other imply a weaker field.

Now let us come to the question of providing a mechanical explanation of electric forces. Suppose the field lines behave like stretched rubber bands and try to shrink in length. If all the field lines between the unlike charges shown in Figure 4.3(a) try to shrink in this way, then

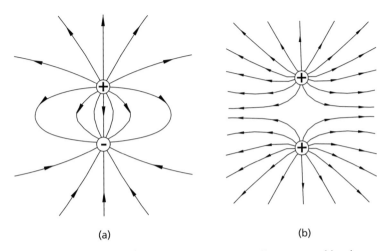

(a)   (b)

Figure 4.3 Electric field lines between (a) two equally strong unlike charges and (b) two equally strong like charges.

they will clearly pull the two charges together. We can thus explain the attraction between unlike charges by postulating that the field lines have a tension and try to shrink along their lengths. Next, let us suppose that nearly parallel field lines lying next to each other push against each other. If that is the case, then the field lines in the central region of Figure 4.3(b) will push each other and, as a result, the two like charges will be pushed away from each other. Thus, if we postulate that the field lines are endowed with the following two properties:

1. All field lines have tension along their lengths and always try to shrink in length;
2. All field lines exert pressure in transverse directions and try to push away nearby field lines that are approximately parallel;

then we arrive at a mechanical explanation of the electric force. Electrical attraction between unlike charges or electrical repulsion between like charges can be thought of as arising out of field lines pulling or pushing against each other.

The above considerations for the electric field can be readily extended to magnetic fields as well. Since electric charges can be separated but not magnetic poles, the discussion is somewhat simpler in the case of electric fields. However, we can understand magnetic forces also by assuming that magnetic field lines have tensions along their lengths and exert pressure in the transverse direction. Things are somewhat different for the gravitational field for which there is only one kind of mass and we always have attraction. We shall not discuss gravitational fields in this book. Einstein had to develop general relativity to treat gravity properly. In the case of electromagnetic fields, which we are interested in, it is possible to develop a mechanical view by assuming that the field lines have the mechanical properties listed above. However, nineteenth-century physicists were not entirely happy with this mechanical view. Since electric and magnetic fields can exist in empty space, they were bothered by the question as to what it is that the field lines must be pulling or pushing in the empty space. They assumed that all of space is filled with the hypothetical elastic medium called ether. The tensions along field lines or pressures perpendicular to them were interpreted as stresses in the ether.

Even a person without formal training in physics would nowadays know that Einstein overthrew the idea of ether when he formulated the special theory of relativity in 1905. Even before this, the theory of ether was already facing insurmountable difficulties. Read Einstein and Infeld's *The Evolution of Physics* to learn more about the fascinating history of this subject. The current view is that we should treat electric and magnetic fields as real entities of which the field lines have the two physical characteristics listed above. We should not invoke some additional medium like the ether to explain these physical characteristics any further. In fact, when dealing with electric and magnetic fields, it is often useful to focus just on their properties without even bothering about the sources (charges, currents or magnets) which produced these fields. We are going to adopt this point of view. We have pointed out that plasmas often have magnetic fields in them. We shall take these magnetic fields as real entities and study their physical characteristics, without bothering too much about the currents which might have produced them.

4.4 Making Little Suns in Our Laboratories

Just after World War II, research on plasmas got a tremendous boost due a very practical consideration. I want to tell you a little bit how this happened, although this will be a slight detour from our main story. Human civilization, as we know it today, depends on a continuous supply of energy. So far much of this energy has come from fossil fuels like coal or oil. At the rate at which we are using up these fossil fuels today, it is estimated that oil will last for only a few decades, whereas coal may last for a few centuries at most. Unless we are able to come up with alternative sources of energy to replace fossil fuels, human civilization will come to a standstill.

Ever since physicists realized that the sun is powered by nuclear reactions, a tantalizing question was whether we could replicate this in our laboratory and generate energy out of nuclear reactions. The atomic bombs dropped on Hiroshima and Nagasaki showed that human beings are capable of using nuclear reactions for terribly destructive purposes. The energy in these atomic bombs was produced by the breaking of heavy nuclei like uranium—a process known as *nuclear fission*. On the other hand, as we discussed in Section 3.1, the energy in the sun is produced by light nuclei combining together, leading to the production of helium from hydrogen. This process is called *nuclear fusion*. The hydrogen bomb developed a few years after World War II is based on nuclear fusion—the same process by which energy is generated in the sun. The big question confronting scientists after World War II was whether nuclear reactions could be made to take place in a controlled way so that we can harness their energy.

We now do have nuclear power plants and some countries like France meet a large part of their energy requirements from nuclear energy. All the present-day nuclear power plants are based on nuclear fission. Heavy nuclei like uranium are broken to provide energy in these plants. A very serious problem of nuclear energy generation in this way is that this process produces radioactive wastes which are difficult to dispose of safely. Secondly, an accident in such a nuclear power plant can release dangerous amounts of radiation. The accidents at Three Mile Island, Cherbonyl and Fukushima Daiichi made the vulnerability of these nuclear power plants very clear to everyone. Finally, the availability of materials which can be used in fission power plants is limited. So, such power plants cannot provide a permanent solution to mankind's energy problem.

If we can generate energy by nuclear fusion—literally producing little suns in our laboratories—then we will have a clean source of energy without the problem of radioactive waste. Also, the supply of materials for such fusion power plants will be virtually inexhaustible because hydrogen is easily available. Any high school chemistry textbook would certainly discuss isotopes. We know that the chemical properties of an atom are essentially determined by the number of protons in its nucleus. If two atomic nuclei have the same number of protons, but different numbers of neutrons, then they would have the same chemical properties and would be known as isotopes of each other. Normal hydrogen has only a proton in its nucleus. But hydrogen has two isotopes: deuterium has one proton and one neutron in its nucleus, whereas tritium has one proton and two neutrons. Now, a helium nucleus has two protons and two neutrons. So you can easily see that two deuterium nuclei combined together would give a helium nucleus. It turns out that the mass of the helium nucleus is less than the combined mass of two deuterium nuclei. So, if we can make two deuterium nuclei combine, a small amount of mass gets converted into energy, providing a supply of energy from such nuclear reactions.

A fraction of the hydrogen atoms in the water molecules of seawater have deuterium as their nuclei instead of a simple proton as is the case in an ordinary hydrogen atom. If we can extract these deuterium nuclei and can combine them in pairs to form helium nuclei, then we can effectively solve the energy problem of the human race. However, we encounter the same problem of electrical repulsion between the deuterium nuclei that we discussed in the context of nuclear reactions in the sun (Section 3.1). If we can make a deuterium plasma with a temperature comparable to the temperature at the centre of the sun, then we would expect some deuterium nuclei to come close together overcoming the electrical repulsion because of their very high velocities and then to take part in nuclear fusion reactions. The main challenge of constructing a nuclear reactor based on this principle is the storage of this very high temperature plasma. Any known material would vaporize at such a high temperature. So we cannot use a container made of any material substance to keep this plasma. Then where do we keep it?

In 1934, when plasma physics was still in its infancy, Willard Bennett made an interesting proposal.[5] He suggested that we could use a magnetic field to keep a plasma confined in a region. Figure 4.4 shows some magnetic field lines around a cylindrical column of plasma. It is not difficult to produce such a magnetic field. One important result that high school students have to learn is that a straight current-carrying wire produces circular magnetic field lines around itself. So, if we send a current through the cylindrical plasma column of Figure 4.4, the magnetic field lines would be exactly as shown. We have pointed out in

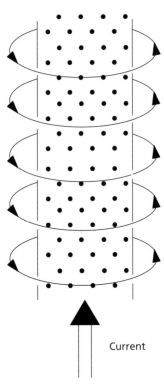

Figure 4.4 A simple possible way of confining a plasma column with the help of magnetic fields. We discuss in the text that this simple mechanism for confining the plasma does not work.

the previous section that magnetic field lines have tension along them and try to shrink in length like stretched rubber bands. Now, a circular magnetic field line can shrink in length only by becoming smaller in radius. As the magnetic field lines try to shrink like this, they exert an inward force on the plasma column, stopping the plasma from spreading out. Bennett thought that this might be a way to keep plasma confined using magnetic fields.

At first sight, Bennett's suggestion seems easy to follow. We heat a column of deuterium plasma to a very high temperature and send a current through it. This current will produce a magnetic field that will keep the plasma confined and nuclear fusion reactions will take place inside it. However, when physicists tried to achieve this in their laboratories, they found that these simple ideas did not work. The plasma kept leaking badly and coming out of the clutches of the magnetic field in various ways. We call these *plasma instabilities*, which made it much more difficult to confine a plasma with magnetic fields than what physicists had initially thought. The plasma behaves like a slippery animal which you are trying to hold with your fingers. The animal just keeps wiggling out!

In the 1950s several countries started research for building plants to generate power by nuclear fusion reactions. Nobody suspected at that time that it would be so difficult to confine plasmas with magnetic fields. It was initially thought that commercial power plants based on nuclear fusion were just around the corner and would be possible in only a few years. Only if nuclear fusion reactions produced more energy than what was used to run the plant, would the power plant be commercially viable. So far, the leaking of plasma has prevented this from being possible.

While the early hope of 1950s of quickly building a commercial power plant based on nuclear fusion was dashed, experts on this subject feel that not everything is lost. Now, based on experience acquired over more than a half a century, plasma physicists have a much better idea of how plasma instabilities may be suppressed and how the plasma can be confined for a longer time. It appears that larger plasma devices are better for confining the plasma. A sufficiently large plasma device is likely to cross the breakeven point and make commercial generation of energy from nuclear fusion feasible. Such a large plasma device has not been built yet, but is in the process of construction in southern France. It has been named International Thermonuclear Experimental Reactor, abbreviated to ITER. This is a very expensive and technologically challenging project involving collaboration amongst several countries. Although the European Union is taking the lead, the other partners are USA, Russia, Japan, India, China and South Korea. ITER is expected to be operational around 2020. Since such a large plasma device had never been built before, nobody can be 100% sure how it will perform. Extrapolations from smaller plasma devices give plasma physicists the hope that ITER will succeed in producing more energy output than what it would require as energy input.

If ITER succeeds (hopefully it will), that will change the course of human history by solving the energy problem of mankind for all time. After taking this detour through an application of plasma physics which

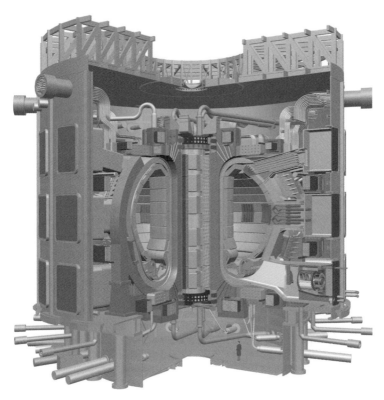

Figure 4.5 The plan for the interior of ITER expected to be ready by 2020. The whole structure will be about as tall as a three-storey building. Credit: ITER Organization.

is likely to transform our lives in very fundamental ways in the not too distant future, we now return to some of the basic scientific issues which will be at the heart of our explanation of the sunspot cycle.

4.5 Coils, Magnetic Flux and Electromagnetic Induction

On 29 August 1831 Michael Faraday, aged 40 at that time, made the most important discovery of his illustrious scientific career—he discovered electromagnetic induction—which, more than any other discovery, made the whole electrical industry possible and which also lies at the heart of almost everything we are going to discuss in this book from now onwards. We know the exact date of this discovery because Faraday had the habit of keeping detailed diaries of his experimental investigations. Oersted's discovery that an electric current (i.e. electricity in motion) gives rise to a magnetic field showed that one can get magnetism out of electricity. Is the opposite possible? Can one get electricity out of magnetism? This is a question which bothered Faraday for many years. Several of his earlier efforts of getting electricity out of magnetism failed, until success came on 29 August 1831.

Let us not discuss Faraday's original experimental setup which was a little complicated. The simplest way of visualizing electromagnetic induction is to think in terms of a coil near which you bring a magnet. You will find that there will be an electric current in the coil whenever the magnet moves with respect to the coil. As we already pointed out, sometimes it is more useful to think in terms of the magnetic field rather than the magnet which produced it. When you move the magnet, you essentially change the magnetic field passing through the coil. Electromagnetic induction basically means that a changing magnetic field can induce an electric current in a coil.

To proceed further, we need to introduce the concept of magnetic flux. It is the amount of magnetic field passing through the coil. For simplicity, we restrict our discussion to a coil with only one turn of wire. Let us consider the simplest case of a uniform magnetic field in a region indicated by straight lines in Figure 4.6(a). We have a coil in this magnetic field. The dashed line in the figure is the projection of this coil in a plane perpendicular to the magnetic field. If you multiply the projected area of the coil inside this dashed line by the magnetic field, you have an estimate of how much magnetic field is passing through the coil. This product is the magnetic flux through the coil in this simple situation. In a more complicated situation where the magnetic field is not uniform, you will have to perform an integration—you will know what I mean if you have some acquaintance with calculus. In Figure 4.6(b) we show another configuration of the coil for which the projected area is less than what it is in Figure 4.6(a) and consequently the magnetic flux passing through the coil is less. In other words, if we rotate the coil in such a way that its configuration changes from what is shown in Figure 4.6(a) to what is shown in Figure 4.6(b), then we have less magnetic flux (i.e. a lesser amount of magnetic field) passing through the coil.

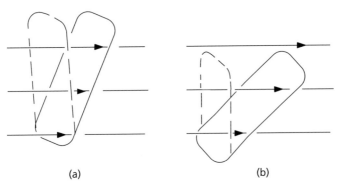

Figure 4.6 An illustration of the concept of magnetic flux through a rectangular coil. (a) and (b) show two configurations of the coil such that the projected area of the coil perpendicular to the magnetic field (indicated by the dashed rectangle) is less in (b) than in (a).

Faraday's law of electromagnetic induction basically tells us that we have a current induced in a coil when the magnetic flux passing through it changes. If the coil shown in Figure 4.6 rotates from the configuration of Figure 4.6(a) to that of Figure 4.6(b), then we shall have a current through the coil while this change is taking place. Now, a current can flow through a coil in two possible opposite directions. In which direction will the current flow? Within two years of Faraday's discovery of electromagnetic induction, Heinrich Lenz discovered a simple law for determining the direction of the current. Let me now try to explain this law. The current induced in the coil will surely produce a magnetic field, as all currents do. There are rules for finding out the direction of the magnetic field produced by a current. But, instead of stating more rules (which are discussed in elementary physics textbooks), let me explain this through pictures. In Figures 4.7(a) and 4.7(b) we see a circular coil (thick line) with currents flowing in opposite directions. The thin lines with circular arrows show the magnetic fields produced by these currents. It is clear that there will be some magnetic flux passing through the coil because of the current. If the magnetic field out of the paper is taken as positive, the magnetic flux through the coil is positive in Figure 4.7(a) and negative in Figure 4.7(b). With this background, we are now ready to state Lenz's law. When the magnetic flux passing through a coil changes, the current is induced in

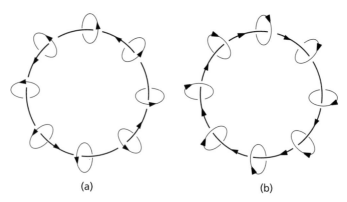

(a)   (b)

Figure 4.7 A circular coil (thick line) carrying a current which produces the magnetic field (thin lines) shown in the figure. (a) and (b) show the two situations of the current flowing in opposite directions.

such a direction that the magnetic flux caused by the current tries to compensate the original change of magnetic flux.

This is a truly remarkable law! If the magnetic flux due to the induced current is capable of completely compensating the original change of magnetic flux, then it is possible that the total magnetic flux through the coil—that is, the sum of the magnetic fluxes due to the external magnetic field and due to the induced current—will not change at all. Can this happen? It seems that this can happen in an idealized limit. One important law of electricity discussed in high school physics classes is Ohm's law of current, named after Georg Ohm. Any coil has some resistance that hinders the flow of current. The lower the resistance, the more easily the current flows through the coil, and compensating for the magnetic flux change by inducing a suitable current will also be easier. If the resistance of a coil is exactly zero, it follows easily from the basic laws of electromagnetism that the magnetic flux passing through such a coil cannot change at all. Readers familiar with calculus and elementary circuit theory may look up a simple proof of this striking result in Appendix C. If the coil is moved in such a way that the magnetic flux due to an external magnetic field changes, then the magnetic flux caused by the induced current will *exactly* negate this. Ordinary coils never have zero resistance and we do not expect this to happen for them. However, if we have a coil with a really miniscule resistance, then the magnetic flux though this coil will remain nearly constant even when the coil is moved around in a magnetic field, the magnetic flux due to the induced current compensating any change in the magnetic flux due to external magnetic fields.

When dealing with a plasma, we do not have discrete coils. However, some of our discussions can be generalized for the plasma, leading to truly far-reaching conclusions. We turn to this issue now.

4.6 Plasmas and Flux Freezing

Water is made up of molecules. But, when we look at water, its granular structure is not apparent to us and we perceive it as a continuous substance. Movements of water can be studied very well with the help of fluid dynamics equations, which assume water to be continuous. In exactly the same way, although a plasma is made of ions and electrons, its movements can often be studied by assuming the plasma to be a continuous substance. The branch of plasma physics in which plasmas are treated in this way is called magnetohydrodynamics or MHD. For the mathematical analysis of sunspots and their cycle, we always treat the plasma in the sun with the help of MHD equations. In our discussion here, we shall now proceed as if a plasma is a continuous substance. In Section 1.4, I mentioned that the central question in our understanding of the sunspot cycle is whether self-excited fluid dynamos are allowed by MHD equations.

Let us begin by considering how the concept of electromagnetic induction, which was first developed for coils, can be applied to a continuous substance like a plasma. Let us consider a ring in the plasma as shown in Figure 4.8. There is no difference between the plasma inside or outside the ring. The surface of the ring is just a hypothetical surface, such that we can treat the plasma inside the ring differently from the surrounding plasma. If there is a magnetic field in the plasma, there will be a certain amount of magnetic flux passing through the ring. Now, if there are movements in the plasma, then the plasma material which made up the ring ABC at time t will move with time. Suppose the material which was at A at time t moves to A' at time t'. Similarly considering all the materials inside the ring, suppose the ring has moved from the position ABC at time t to the position $A'B'C'$ at time t'. We can now treat the ring like a coil and apply the principles of electromagnetic induction to it. As the ring moves, the magnetic flux passing through it may tend to change, inducing currents which will try to

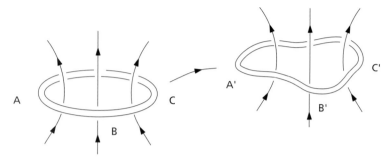

Figure 4.8 A ring inside a plasma which has moved from the position ABC to the position $A'B'C'$ in an interval of time. The magnetic fluxes associated with the ring are indicated.

negate this change of magnetic flux. In the case of a coil, we sometimes distinguish between the flux due to the external magnetic field and the flux due to the magnetic field of the current induced in the coil. When considering the ring in the plasma, such distinctions are no longer meaningful because all magnetic fields are produced by currents flowing through some region of the plasma. We can only talk of the net magnetic flux arising out of all possible currents in the plasma. Now, consider what happens to this magnetic flux when the ring ABC moves with the movements inside the plasma. In the case of the coil, we know that the net magnetic flux through the coil will not change at all if the resistance of the coil is negligible. Following the same considerations, we expect that the magnetic flux through the ring in Figure 4.8 will not change when the ring changes its position, provided the effect of resistance is small.

Now our job is to figure out the effect of resistance in a ring inside the plasma. One needs to analyse the full mathematical equations to address this question. Here I give an idea of how things go. I have already mentioned that Maxwell captured the essence of electromagnetism in a set of elegant equations. One of Maxwell's equations is nothing but the law of electromagnetic induction written in an advanced mathematical notation. To study the effect of resistance, one also needs Ohm's law in addition to Maxwell's equations. The simple form of Ohm's law that we learn in high school is mainly applicable to coils. However, it is possible to generalize Ohm's law for a continuous medium like a plasma. By combining Ohm's law with Maxwell's equations, we get an equation known as the *induction equation*, the central equation of MHD which describes how magnetic fields in the plasma would change with time. It indeed turns out from the induction equation that the magnetic flux through a ring as shown in Figure 4.8 remains constant as it moves, if the effect of resistance is small. Readers who are comfortable with vector analysis and who are curious to know what the basic equations look like may turn to Appendices D and E.

We now need to explain what we mean by the effect of resistance being small. Consider two copper wires of the same length—one very thin and the other quite thick. We know that the thick wire has less resistance because the current has more space to flow through the thick wire. In the case of the plasma also, if the plasma has a larger volume and the current has more space to flow through, the effect of resistance is less. In other words, when we consider plasma systems of larger and larger size, the effect of resistance becomes more negligible and the magnetic flux through a moving ring in that plasma system should remain constant to a greater extent. To be more precise, for a plasma system, one can calculate a number called the *magnetic Reynolds number*.[6] This number arises quite naturally out of the mathematical equations (see Appendix E). Here I shall only say that this number is usually much smaller than 1 for laboratory plasma systems and much larger than 1 for astrophysical plasma systems. Basically, the larger the size of the system, the larger is the magnetic Reynolds number. Now, the mathematical equations tell us that the effect of resistance becomes increasingly negligible for systems with larger and larger magnetic Reynolds numbers. For astrophysical plasma systems with large magnetic Reynolds numbers, one can forget the effect of resistance and the magnetic flux through a moving ring inside the plasma would essentially remain constant. This is perhaps the most important result in the theory of astrophysical plasmas and is known as *Alfvén's theorem of flux freezing*, after the Swedish scientist Hannes Alfvén who arrived at this theorem in 1942.[7] For this theorem and other important contributions in MHD, Alfvén was awarded the Nobel Prize in 1970.

Do not worry if you are not feeling fully comfortable with this theorem. It takes time to grasp its full significance. Let us consider an example which may help you. Suppose there is a chunk of plasma between the pole pieces of a magnet as shown in Figure 4.10(a). We now move the chunk of plasma away from the pole pieces. What will happen to the magnetic field? In a laboratory situation where the

Figure 4.9 Hannes Alfvén (1908–1995), who made fundamental contributions to MHD and won the 1970 Physics Nobel Prize for these contributions.

magnetic Reynolds number is low, nothing unusual will happen. The magnetic field will remain between the pole pieces and the chunk of plasma will be out of the magnetic field, as shown in Figure 4.10(b). Now suppose that our system is of astronomical size and the magnetic Reynolds number is very high. Then we can imagine a ring inside the chunk of plasma and we know that the magnetic flux passing through this ring will have to remain constant. This can happen only if the magnetic field takes up the configuration shown in Figure 4.10(c). In other words, as the chunk of plasma is moved, Alfvén's theorem of flux freezing requires that the magnetic field should get drawn out with the plasma. We say that the magnetic field is frozen in the plasma and moves with it. By moving the chunk of plasma further and further, we can make the magnetic field lines more and more bent. We thus see that, if the magnetic Reynolds number is high, we can bend a magnetic field by suitable plasma motions, as if the magnetic field is some plastic

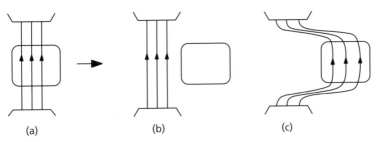

Figure 4.10 (a) A chunk of plasma inside a magnetic field between the pole pieces of a magnet. (b) and (c) show what will happen to the magnetic field when the chunk of plasma is moved outward in the situations of low magnetic Reynolds number and high magnetic Reynolds number respectively.

material. This is totally unlike anything we see in laboratory situations where the magnetic Reynolds number is low.

Learning new things may be difficult. But sometimes unlearning old things is even more difficult. Physics courses at high school and college levels teach many things about magnetic fields. Any physics student develops some intuition into how magnetic fields behave under various laboratory situations. The initial difficulty a student faces in learning about magnetic fields in astronomical settings is that the old intuition based on high school and college physics courses often goes against what he or she now has to learn. Since magnetic fields behave so differently in laboratory and astronomical settings (where the sizes of the systems are vastly different), Alfvén coined the term cosmical electrodynamics for the study of electromagnetic phenomena in the cosmical scale. In fact, in 1950 Alfvén wrote a famous book titled *Cosmical Electrodynamics*, which was very influential in the development of this subject.

In a laboratory situation, we usually consider the current to be more basic than the magnetic field which is produced by the current. For a given current, we can calculate the magnetic field by applying the basic laws of electromagnetism, such as Ampere's equation. The magnetic field is there as long as the current flows and disappears when the current is switched off. The situation is completely different in astronomical settings. We have pointed out that the effect of resistance is very small in large astronomical systems. So currents and magnetic fields can persist in astronomical systems for a very long time, without undergoing significant attenuation due to resistance. Magnetic fields seem to acquire some independent life of their own in astronomical settings. They move with the moving plasma and can be bent or twisted by appropriate plasma movements, like a plastic material. In astronomical settings, it is useful to regard the magnetic field as more basic than the current. Currents simply adjust themselves in such a way that the flux freezing condition for any arbitrary imaginary ring inside the plasma is satisfied. Once you have mastered the concept of flux freezing, you can understand the behaviour of the magnetic field more easily than that of the current. The view that the magnetic field is more basic than the current in situations of high magnetic Reynolds number emerges readily out of Alfvén's pioneering investigations. However, ironically, Alfvén himself opposed this view in later life. He started arguing that one should regard the current as more fundamental even in astronomical settings, negating much of the insight which his own earlier work had provided! Much of the work he did in later life was controversial and was even considered wrong by other scientists. Although Alfvén's early work had a tremendous impact on the study of astronomical magnetic fields, his later work was ignored by most of the younger scientists working in this field, with many of whom Alfvén had famously frosty personal relationships. In spite of being a Nobel Prize winner, Alfvén in his later life sometimes even had difficulties in publishing his papers. Editors of reputed journals often refused to publish his papers. When I started my research in this field in the 1980s, I came across so many contradictory accounts of Alfvén—from extremely high praise to total disdain—that I was terribly curious to meet the man. Unfortunately I never had an opportunity of meeting Alfvén.

In the following chapters, I shall assume that the sun's magnetic field can be best understood if we regard the magnetic field as more basic than the current. We shall also repeatedly use the ideas introduced a few pages earlier that magnetic fields have tensions along them and exert pressure sideways. I mentioned that initially there were some efforts to interpret these tensions and pressures as stresses in the ether. This programme had to be abandoned as ether fell out of favour in scientific circles. If we have magnetic fields in the plasma, then these tensions and pressures can be viewed as stresses in the plasma. A magnetic field, which acquires the characteristics of a plastic material when embedded in a plasma, imparts to the plasma in turn some characteristics which the ether was once assumed to possess.

Whenever we derive an important theoretical result in science, we like to verify it experimentally to ensure that our understanding is correct. Some of the most intriguing results of MHD are obtained for situations of high magnetic Reynolds number. They cannot be verified in our laboratories where the magnetic Reynolds number is usually much lower. The sun is the nearest large plasma body in which we can look for some of the effects our mathematical equations suggest. We often say that the sun is our best laboratory for MHD. There is indeed a symbiotic relation between MHD and the study of the sun. On the one hand, many phenomena occurring in the sun require MHD for their theoretical explanation, as we shall see in the subsequent chapters. On the other hand, the sun is the best place for verifying many theoretical predictions of MHD.

4.7 A Part of the Central Dogma Explained

In Section 2.5, I had introduced what I called the central dogma of solar dynamo theory. Look at Figure 2.10 to remind yourself what is a toroidal magnetic field and what is a poloidal magnetic field. One part of the central dogma is that the poloidal field should give rise to the toroidal field. Armed with Alfvén's theorem of flux freezing, I am now ready to tell you how this happens. In fact, what I am now going to tell you is a very obvious extrapolation of what you see in Figure 4.10(c).

We have discussed in Section 2.1 how Carrington discovered that the regions of the sun near the equator rotate faster than the regions in higher latitudes. Then, in Section 3.7, I have pointed out how helioseismology mapped the rotation rate throughout the sun's interior. Now we come to the question: if there is a poloidal magnetic field inside the sun, what will be the effect of the differentially rotating sun on this poloidal magnetic field? As sketched in Figure 4.11, Alfvén's theorem of flux freezing provides us with a clue to answering this question. Figure 4.11(a) shows a poloidal magnetic field line passing through the sun, within which the equatorial region is rotating faster. Just as the chunk of plasma in Figure 4.10(c) dragged the magnetic field lines out with it, we would expect the faster moving plasma near the sun's equator to drag out the poloidal field line, as shown in Figure 4.11(b). On looking at this resulting magnetic field line in Figure 4.11(b) carefully, you can easily see that it is now no longer an entirely poloidal magnetic field line. A part of the magnetic field line is now a toroidal magnetic

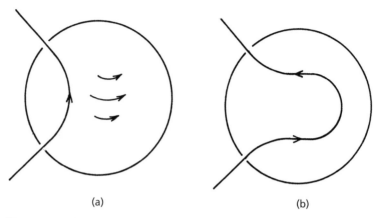

(a)   (b)

Figure 4.11 The production of toroidal magnetic field from poloidal magnetic field in the sun. (a) An initial poloidal field line, with small arrows indicating rotation varying with latitude. (b) A sketch of the field line after it has been stretched by the faster rotation near the equatorial region.

field. In other words, we have got a toroidal magnetic field starting from a poloidal magnetic field, as required by the central dogma.

You can see in Figure 4.11(b) that the toroidal magnetic field which has resulted from the initial purely poloidal magnetic field of Figure 4.11(a) has opposite directions in the two hemispheres—exactly like the toroidal field sketched in Figure 2.10(a). While discussing the central dogma in Section 2.5, I pointed out that sunspot pairs are believed to be produced from the toroidal magnetic field. We shall discuss in the next chapter why parts of the toroidal magnetic field may float up to the surface from the sun's interior to form sunspot pairs, as sketched in Figure 2.8(a). Since the toroidal magnetic field in Figure 4.11(b) has opposite signs in the two hemispheres, it is easy to see that the sunspot pairs which result from it will have opposite polarities in the two hemispheres, in conformity with the observational data seen in the magnetogram map of Figure 2.7. The other crucial part of the central dogma is that the toroidal magnetic field should give rise to the poloidal magnetic field. How this happens is the question which lies at the heart of dynamo theory. In Chapters 6 and 7, I shall introduce the crucial theoretical ideas that explain how the poloidal magnetic field arises out of the toroidal magnetic field. We shall see that Alfvén's theorem of flux freezing will again be the central concept behind these theoretical ideas.

The phenomenon of electromagnetic induction, on which Alfvén's theorem of flux freezing is based, is behind the entire central dogma of solar dynamo theory, just as it is behind the whole of the electrical industry.

4.8 A Turning Point in Our Journey

We have now reached a turning point in this book. We have discussed many aspects of sunspots and the sunspot cycle in the first two chapters. Then Chapter 3 was devoted to discussing what the interior of the sun is like and in this chapter we have introduced some basic ideas about plasmas. Only within the previous couple of pages, have I at last started providing explanations for some aspects of our central dogma. Now the real explaining will begin from the next chapter. We shall see how various things we have already discussed can be put together, providing explanations for different aspects of sunspots. So far we have been casting our net wide. Now the job of hauling the catch begins.

Apart from Alfvén's theorem of flux freezing, the fact discussed in Section 4.3 that magnetic fields have tensions along field lines and exert sideways pressure will be needed in explaining many phenomena connected with sunspots. I would urge readers to keep these important concepts from this chapter always in mind while reading the remainder of the book. For mathematically sophisticated readers, Appendix F gives an idea of how the concepts of magnetic tension along field lines and magnetic pressure sideways arise out of the basic mathematical equations.

5

Floating Magnetic Buoys

5.1 Boiling a Magnetic Field

Take another careful look at Figure 1.3 showing two sunspots on the surface of the sun. The fact that sunspots are regions of concentrated magnetic field should by now be firmly imprinted in your mind, since I have repeated this so many times. Looking at Figure 1.3, you may be tempted to think that a sunspot is some kind of a magnetic saucer floating on the sun's surface. However, we have discussed the principle that magnetic field lines cannot end, which led us to understand why new poles appear in the cleavage region when you break a bar magnet (see Figure 2.5) or why isolated magnetic poles do not exist (see Figure 2.6). This principle should make it clear that a sunspot cannot be a shallow magnetic saucer, since the magnetic field lines cannot suddenly end below the sunspot. Magnetogram maps like the one shown in Figure 2.7 suggest that the magnetic configuration below a sunspot pair may be as shown in Figure 2.8(a). In other words, we think that a bipolar sunspot pair is produced by a part of a toroidal magnet field rising up to the surface. Towards the end of the previous chapter, we have used Alfvén's theorem of flux freezing to show that the varying rotation rate (or the differential rotation) inside the sun would tend to produce a strong toroidal field, as explained through Figure 4.11. Now we have to explain why a part of this toroidal field may float up to the sun's surface to produce the bipolar sunspot pair. This is the main subject of this chapter.

Before we come to this question, I want to dispose of a couple of other questions which may arise in your mind. You find that a bar magnet produces a magnetic field all around it. A magnetic compass needle brought at any point near the bar magnet will show a deflection due to the magnetic field. On the other hand, magnetic fields of sunspots are concentrated within them with very little magnetic field outside. I should mention here that modern research in the last few years has

shown that the sun's surface outside of sunspots is not completely devoid of magnetic fields. We shall discuss this point in Chapter 7. For the
time being, let us ignore this extra complication. It is certainly technically correct to say that sunspots are the most prominent and largest
regions of magnetic field concentration on the sun's surface. This brings
us to our first question: why is the magnetic field at the sun's surface
concentrated inside sunspots, unlike the magnetic field of a bar magnetic which fills up all space around it? A second related question is the
following: why are sunspots, which are regions of concentrated magnetic field, darker than the surrounding regions of the sun's surface?
We shall see that these two questions are closely related to each other.
While seeking the answer to the first question, we shall also find an
answer to the second.

We have discussed in Section 3.3 what the interior of the sun is believed to be like. I pointed out that heat is transported by convection
in the outer region of the sun, from a depth of 200,000 kilometres to
the surface. Look at Figure 3.3 to refresh your memory. A sunspot is essentially a magnetic structure sitting in the midst of convecting plasma
within which hot materials are going up and cold materials are coming
down. The granular pattern around the sunspots in Figure 1.3 shows
convection in action. To understand why we have darkened regions of
concentrated magnetic field that we call sunspots, we need to study
the interaction between the magnetic field and convection. This subject
is called *magnetoconvection*. The first important studies in this field were
carried out by Chandrasekhar.

I have written about the famous Eddington–Chandrasekhar controversy around dying stars in Section 3.4. I also mentioned that Chandrasekhar wrote a monograph on the subject of stellar structure and then
moved on to do research in other areas of astrophysics. From then onwards, one sees a pattern in Chandrasekhar's career. He would work in
a particular area of astrophysics for a few years, write a monograph on
it and then leave that area completely to work in another area. During the 1950s Chandrasekhar worked on MHD, which culminated in
his publishing in 1961 a monograph with a rather technical-sounding
title *Hydrodynamic and Hydromagnetic Stability*. It gives an introduction to
magnetoconvection, amongst many other important topics.

I now have to refer back to our discussion in Section 3.2 describing
how Schwarzschild discovered the condition of convection inside stars.
If the temperature difference between the bottom and the top of a gas in

a gravitational field is larger than a critical value—or, in slightly more technical language, if the vertical temperature gradient inside the gas is stronger than a critical value—then heat is transported by convection inside the gas. Schwarzschild did his calculation for a gas obeying the ideal gas law. We also see convection in a liquid like water when it is heated in a pan to make a cup of tea. The calculation for convection in a liquid is a little bit more difficult than that for a gas, although the basic physical picture is the same. When water at the bottom of the pan becomes hotter, it expands and becomes lighter than the water at the top. The hotter lighter water rises and the colder heavier water from the top sinks, setting up convection. The calculation for this type of convection was done by Lord Rayleigh in 1916—about a decade after Schwarzschild's calculation for convection in gases. Rayleigh received the Nobel Prize for the discovery of argon, but also made important contributions to theoretical physics. Rayleigh found a condition for convection in a liquid somewhat reminiscent of what Schwarzschild had found for gases. When the vertical temperature gradient in a liquid is stronger than a critical value, Rayleigh showed that convection would take place.

Chandrasekhar extended Rayleigh's calculation for a special kind of liquid: a plasma with a vertical magnetic field. Figure 5.1(a) shows a schematic sketch of this. Suppose we have a plasma behaving like a liquid, with a vertical magnetic field, and we heat it from below. Since the liquid state is the second state of matter and we have called the

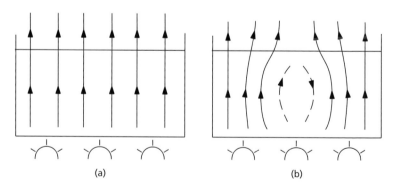

(a)   (b)

Figure 5.1 An electrically conducting liquid with a vertical magnetic field heated from below. (a) The initial configuration. (b) The configuration after convection starts, the dashed lines indicating movements of the liquid.

plasma state the fourth state, you may wonder whether something being a liquid and a plasma at the same time is a contradiction in terms. I have pointed out in Section 4.2 that the main difference between an ordinary gas and a plasma is that a plasma is a good conductor of electricity (see Figure 4.2). If a liquid is a good conductor of electricity, then it also would have some characteristics of a plasma and would obey the basic equations of MHD. Mercury is one such liquid. In the early years of MHD research, many experiments were done with mercury to verify the predictions of MHD equations. In fact, several groups carried out experiments with mercury to verify Chandrasekhar's theory of magnetoconvection.

Let us now discuss what happens to the system shown in Figure 5.1(a). If you have ever watched a pan of boiling water carefully, you would know that convection involves disorderly motions of hot liquid going up and cold liquid coming down. If the magnetic field moves with the plasma—at least to some extent—as we expect from Alfvén's theorem of flux freezing, then the magnetic field will also get disorderly. Now, when we discussed the reality of fields in Section 4.3, we concluded that magnetic field lines have tension along their length. Because of this tension, magnetic field lines resist being disorderly. As a result, it is more difficult to start convection in a liquid like mercury when a magnetic field is present. Chandrasekhar showed in 1952 that the temperature gradient needed to start convection in the presence of a magnetic field is stronger than that needed to start convection in its absence.[1] This result was beautifully confirmed by mercury experiments.

While a magnetic field may make it more difficult for convection to start inside a liquid, what would be the nature of convection when it eventually starts after making the temperature gradient sufficiently high? This is a question which can be answered only by doing numerical simulations on a computer. Since we shall have to talk about many numerical simulations from now onwards, I shall soon say a few words as to what numerical simulations are. But, before that, let me tell you what we can learn about magnetoconvection from numerical simulations. Nigel Weiss carried out the first pioneering numerical simulations of magnetoconvection. What follows is based on the insight provided by the numerical simulations of Weiss, his co-workers and other groups.

When the temperature gradient is high enough, convection starts, but magnetic tension tends to oppose it. As a result, nature finds it

Figure 5.2 Nigel Weiss (1936–), whose pioneering numerical simulations showed how magnetic flux tubes form in the sun's convection zone.

expedient to divide the system neatly into two kinds of regions. In some regions, the magnetic field lines (shown by solid lines in Figure 5.1(b)) get bundled up and their tension prevents convection from taking place. In other regions from where magnetic field lines have been pushed aside, convection can take place freely, as shown in Figure 5.1(b) by dashed lines indicating the movements of the liquid. The bundles of magnetic field lines are called *magnetic flux tubes*. Since convection is suppressed inside these, heat transport becomes less efficient and numerical simulations show that the tops of these flux tubes have lower temperatures than the surrounding surface. Numerical simulations of magnetoconvection thus give us some idea about why we have objects like sunspots having concentrated magnetic fields with lowered temperature. Magnetic fields in the convection zone are likely to be concentrated into flux tubes due to interaction with convection. A sunspot is presumably such a vertical flux tube in which heat transport is less efficient due to the suppression of convection, making the sunspot colder and darker than the normal regions of the sun's surface. Weiss's first simulation in 1966 demonstrated the formation of magnetic flux tubes.[2] However, I should mention that in 1941, several decades before the numerical simulations of magnetoconvection and even a decade before Chandrasekhar's work on this subject, Ludwig Biermann already argued on the basis of his remarkable physical intuition that convection would be suppressed inside sunspots due to magnetic tension and that is why sunspots are colder.

5.2 Simulating the Physical World on Our Computers

Perhaps this is an appropriate place for me to say a few words about numerical simulation, since we shall have to discuss many simulations in the coming pages of this book. A typical law of physics often relates the rate of change of a quantity to something else. One example of such a law is Newton's second law of motion, which lies at the foundation of the entire science of mechanics. According to this law, the force acting on a particle has to be equal to the rate of change of its momentum. Now, calculus is the branch of mathematics which is exactly meant for tackling rates of change of various things (Newton himself being one of the inventors of calculus). A law of physics dealing with the rate of change of something can usually be translated into a mathematical equation of calculus. This type of equation is called a differential equation. In an elementary college course on calculus, one learns many rules for solving differential equations. But we do not have rules for solving all kinds of differential equations. Suppose we come across a particular differential equation which cannot be solved by applying the rules given in the college calculus textbooks. What do we do with such a differential equation?

Let us come back to Newton's second law of motion that the force is equal to the rate of change of momentum. Since the momentum is the product of mass and velocity, for a particle of constant mass, the force will essentially produce a rate of change of velocity. Now the velocity is the rate of change of position. So the force should cause what would be like a second-order rate of change of position. This can be expressed in the form of a suitable differential equation. Now, very often, we want to find out how the position of the particle changes with time. We should be able to get this information by solving the differential equation. Can our differential equation be solved with the help of the rules one finds in college calculus textbooks? Consider the situation of Galileo dropping a ball from the leaning tower of Pisa, as he was supposed to have done to disprove Aristotle. The force due to the earth's gravity acting on the ball is a constant force in this situation. When the force is constant, the differential equation describing the motion can easily be solved. The solution tells us that the distance through which the ball falls in a certain time is proportional to the square of that time. In other words, in 2 seconds the ball will fall through 4 times the distance it has fallen through in 1 second, in 3 seconds it will fall through 9 times the distance it has fallen through in 1 second, and so on. It turns out that the strength of the earth's gravity is such that the ball falls through 4.9 metres in 1 second. So we can very easily tell the distance through which the ball would have fallen in t seconds. It is simply $4.9t^2$ metres.

Now consider the situation where the force acting on a particle is not constant, but varies with position and time in a complicated way. For this general case, the rules given in a college calculus textbook may not enable one to write down the solution of the differential equation. If we want to find how the position of the particle will keep changing with time, we have to solve the differential equation numerically. If we know the position and the velocity of the particle at one instant of time, a little bit of numerical work using the differential equation will give the position and the velocity of the particle at a slightly later instant of time. Once we know the position and the velocity of the particle at this slightly later instant of time, the differential equation will again give us the position and the velocity of the particle at a still slightly later instant of time. Proceeding in this way, we can find out how the position and the velocity of the particle will keep changing with time. Nowadays one usually writes a computer code to do such calculations on a computer. This is the essence of numerical simulation. If we know the correct differential equation describing a physical system, then the output of the computer code for solving this differential equation will tell us how the physical system will change with time. Here I should mention that writing a code to solve a complicated differential equation can sometimes be an extremely challenging and time-consuming task. It can easily take several months to write a computer code to solve the equations of MHD in a complicated situation!

Let us now come to the question of how we do a numerical simulation of magnetoconvection. We can take an electrically conducting liquid with a vertical magnetic field as shown in Figure 5.1(a) as our initial condition. If we put a temperature gradient sufficient to start convection, then motions will get induced within this liquid. The equations of MHD would tell us how the motions and the magnetic field will keep changing with time. So we have to write a computer code to solve these equations. When we run this code on a computer, the computer tells us how the motions and the magnetic field would change with time. If we keep running the computer code, we would find that the configuration of the magnetic field changes from what is shown in Figure 5.1(a) to what is shown in Figure 5.1(b).

Large-scale electronic computers started being developed from the time of World War II. Electronic Numerical Integrator And Computer (abbreviated as ENIAC) developed in the University of Pennsylvania in 1946 is usually regarded as the first general-purpose electronic computer. During the 1950s and 1960s, electronic computers started becoming available to professors in many universities to carry out their research. Up to that time, only very limited progress was possible in problems in which the basic differential equations could not be solved using the rules in calculus textbooks. The availability of electronic computers revolutionized almost all areas of physics including astrophysics and allowed physicists to investigate problems which were not possible to investigate till that time. When Weiss wrote a computer code in 1966 to solve the MHD equations and showed how magnetic flux tubes form, it was one of the first landmark numerical simulations in the history of theoretical solar physics.

Let me end this section by mentioning that I am personally indebted in many ways to Nigel Weiss, who had been a professor at Cambridge University for many years until his retirement a few years ago. Nigel once visited Parker in Chicago when I was a PhD student there. From that time onwards, Nigel has always taken a keen personal interest in my research and in my career. When I decided to return to India to establish a research group, I went through a very difficult initial phase which I shall describe later in this chapter. Nigel is one senior person in our field who always kept track of what I was doing and encouraged me in various ways. Nigel and I hold rather different viewpoints on some important aspects of solar dynamo, as I shall discuss later. Whenever we have met, we have vigorously argued about these. While neither of us might have succeeded in converting the other to his viewpoint, scientific discussions with Nigel have always been extremely instructive and stimulating for me. I think of Nigel as my second guru after Gene Parker.

5.3 Magnetic Buoyancy

In the year 1955 in which Parker wrote his fundamental paper on dynamo theory, he made another tremendously important contribution to solar physics in another paper.[3] A part of the toroidal magnetic field in the sun's interior has to float up to the sun's surface to produce a bipolar sunspot pair, as shown in Figure 2.8(a). Parker invoked a 2300-year-old idea due to Archimedes to explain why the toroidal magnetic field would float up.

We believe that magnetic fields exist in the form of magnetic flux tubes throughout the sun's convection zone due to interactions with convection. Figure 4.11 shows how any magnetic field in the interior of the sun would be stretched by the differential rotation to produce a strong toroidal magnetic field. This toroidal magnetic field is expected to exist in the form of toroidal magnetic flux tubes due to interactions with convection. In such a flux tube, if the pressure inside is more than the external pressure, then the cross-section of the tube will increase. On the other hand, if the pressure inside this flux tube is less than the external pressure, then the flux tube will shrink. So we would expect a pressure balance such that the pressures inside and outside are the same, ensuring that the cross-section of the flux tube does not expand or contract. Now, while discussing the reality of fields in Section 4.3, I have pointed out that a magnetic field exerts a sideways pressure. Inside the magnetic flux tube, we have both the usual gas pressure and this magnetic pressure. On the outside, however, we do not have significant magnetic field and we basically have only gas pressure. Hence we expect an equation of the form

External gas pressure = Internal gas pressure + Magnetic pressure.

Such an equation can hold only if the internal gas pressure is less than the external gas pressure. The ideal gas law (discussed in Appendix A) tells us that one way of making the pressure less (though not the only way) is to make the density less. We may therefore expect that the internal gas density of the magnetic flux tube will often be less than the external gas density. Archimedes taught us in the third century BC that, if an object immersed in a fluid has less density than the surrounding fluid, then it will become buoyant and rise up against the gravitational field. Parker realized that portions of the toroidal magnetic flux tube having density less than the surrounding density would be buoyant and would rise up. Since the magnetic field should be frozen in the plasma due to Alfvén's theorem, the plasma and the magnetic field would rise as one entity to produce a bipolar sunspot pair as shown in Figure 2.8(a). Parker called this *magnetic buoyancy*. A short technical introduction to magnetic buoyancy is given in Appendix G.

The idea of magnetic buoyancy is a simple, elegant and powerful idea. Since the toroidal magnetic flux tubes in the two hemispheres are expected to have oppositely directed magnetic fields, as should be clear from Figure 4.11(b), parts of these toroidal magnetic flux tubes rising due to magnetic buoyancy would explain the distribution of magnetic fields in the magnetogram map of Figure 2.7. However, one important question still remains. Why do only parts of the toroidal magnetic flux tubes become buoyant and rise, while other parts remain inside the sun? To address this question, we first have to find out in which regions inside the sun the toroidal flux tubes form.

In the next two chapters, we shall discuss how the poloidal magnetic field of the sun arises. If there are any poloidal magnetic fields inside the sun, they are expected to be stretched by differential rotation to produce the toroidal magnetic field as sketched in Figure 4.11. We expect the strongest toroidal magnetic fields to be produced in the regions where differential rotation is the strongest. We now turn to Figure 3.9 to judge where the differential rotation is the strongest. This would be indicated by a crowding of the contours in the figure. We note that this happens at the bottom of the convection zone. In the discussion accompanying Figure 3.9, I already mentioned that this layer of concentrated differential rotation is called the tachocline. We expect the toroidal magnetic flux tubes to be primarily produced in the tachocline located at the bottom of the sun's convection zone.

While analysing magnetic buoyancy, Parker arrived at a very important conclusion. He found that, within the convection zone where the Schwarzschild condition for convection is satisfied, the magnetic buoyancy turns out to be particularly strong, with convection and magnetic buoyancy reinforcing each other. On the other hand, below the bottom of the convection zone where the Schwarzschild condition for convection is not satisfied, the magnetic buoyancy gets suppressed to a large extent. Let us consider a toroidal magnetic flux tube produced at the bottom of the convection zone. There are always some disturbances present in the convection zone. Because of these disturbances, it may happen that a part of the magnetic flux tube is within the convection

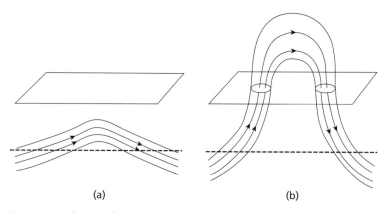

(a)   (b)

Figure 5.3 The rise of a magnetic flux tube through the sun's convection zone. (a) The initial configuration showing a part of the flux tube above the bottom of the convection zone (indicated by the dashed line). (b) The final configuration after the part of the flux tube inside the convection zone has risen through the surface.

zone, whereas other parts are below the bottom of the convection zone. This is shown in Figure 5.3(a) where the dashed line indicates the bottom of the convection zone. Magnetic buoyancy will be very strong within the part of the flux tube inside the convection zone, above the broken line. This part will rise rapidly. On the other hand, magnetic buoyancy will get suppressed to a large extent in the parts below the bottom of the convection zone. These parts of the flux tube will remain anchored at the bottom of the convection zone. Figure 5.3(b) shows the final configuration of the magnetic flux tube after the part originally within the convection zone has risen through the surface, piercing the surface in two regions which will now become the two sunspots with opposite magnetic polarities. Sunspots are basically floating magnetic buoys tethered at the bottom of the sun's convection zone.

Now look again at Figure 2.8(a). To understand in detail how sunspots form, we need to start from an idealized initial configuration of a toroidal magnetic flux ring going around the sun's rotation axis at the bottom of the convection zone. A part of this flux ring may be assumed to lie within the convection zone, while the remaining parts of the flux ring would be below the bottom of the convection zone. This flux ring would appear as shown in Figure 2.8(a), except that the part which has come out should initially be only slightly above the bottom of the convection zone. Eventually this part would rise due to magnetic buoyancy to produce the magnetic configuration shown in Figure 2.8(a). How this rise occurs has been investigated through numerical simulations. I and my co-workers were the among the first to carry out such simulations.

5.4 A Personal Note

At this stage, I would ask the reader's indulgence and would first describe how I got into this field, before coming to my work on magnetic buoyancy. When I was growing up, India was a desperately poor country and opportunities for scientific research were limited. Not that India did not have her scientific heroes! Even when India was under British rule and conditions were even more difficult, some Indian physicists reached the pinnacles of glory. In Section 4.2, I discussed the ionization equation of Meghnad Saha, who came from among the poorest of the poor in India. Saha's college classmate S. N. Bose formulated what is called the Bose–Einstein statistics and created a new branch of physics—quantum statistical mechanics. C. V. Raman, who was slightly older than Bose and Saha, was the first Asian to win a Nobel Prize in physics for his discovery of Raman scattering. Certainly Raman, Bose and Saha were inspirational figures for us. In spite of the difficulties they had to overcome, they worked at a time when physics all over the world was done with limited facilities and limited funds. Competing internationally at the cutting-edge research frontier from an isolated place was perhaps more possible at that time. In any case, Raman, Bose and Saha were all dead by the time I was an undergraduate student. As undergraduate students in India, we never witnessed the creative process of important new physics being created before our eyes. Those of us who studied in Presidency College in Kolkata had the good fortune of having a living legend as our professor: the general relativist Amal Raychaudhuri. The equation he came up with for his doctoral dissertation is now known as the Raychaudhuri equation and provided the starting point for the singularity theorems of Hawking and Penrose. We hero-worshipped Raychaudhuri for his extraordinary physics teaching, which inspired a generation of physics students in Kolkata to take up theoretical physics for their career. Ashoke Sen, India's most distinguished theoretical physicist of our generation, was two years senior to me in Presidency College. A few years later, when I myself had to teach physics to my

Figure 5.4 A view of the Laboratory for Astrophysics and Space Research, as it was when I was working there. The right corner room on the upper floor was Chandrasekhar's office, whereas the left corner room on the upper floor was Parker's office. The room next to Parker's office was my office.

students, I had Raychaudhuri as my model and consciously tried to imitate his teaching style. Professor Raychaudhuri often recounted to us the bitter experiences he had as a young researcher in a stifling environment of mediocrity. He encouraged us to try for admission to places that would offer a stimulating intellectual environment. After completion of my undergraduate studies in 1980, I accepted the fellowship offered by the University of Chicago to do my PhD there.

One of the courses in our first-year curriculum at the University of Chicago was 'Plasma Astrophysics' taught by Gene Parker. I was completely bowled over by the beauty of the subject. I felt that this is a subject compatible with my taste and temperament. I started thinking about the possibility of doing my PhD thesis on this subject under the guidance of Parker. Some senior astronomy students cautioned me that, although Parker was an extraordinary scientist and extraordinary teacher, he was not known to be a good research supervisor. He apparently never 'guided' his PhD students. If I worked with him, I was told that I would be left to fend for myself. Being very young and inexperienced at that time, I did not fully understand the significance of what these senior students told me. I was happy when Parker agreed to accept me as his PhD student.

In those days, most of the astronomy faculty members and students in Chicago worked in a building called Astronomy and Astrophysics Center. The building next to it—Laboratory for Astrophysics and Space

Research—mainly housed the large cosmic ray research group. However, two of the most beautiful offices in that building were given to two theoretical astrophysicists: Chandrasekhar and Parker. Chandrasekhar was past 70 at that time and had stopped taking students after a heart attack a few years earlier. Although he was still working as hard on his research projects as a young man, there were no students working with him when I was in Chicago. After my senior Tom Bogdan completed his PhD with Parker, for a while I was the only theory student having a very nice office in the building Laboratory for Astrophysics and Space Research. Parker's office was next to mine on one side and Chandrasekhar's office was on the other side three or four doors down the corridor. While I was attending the first scientific meeting of my life—a meeting of the American Astronomical Society—somebody asked me at the dinner table: 'How big is your theory group?' I replied: 'It is a very small theory group with only three members.' The next question was: 'Who are the members?' I casually said: 'Oh, besides myself, the other two members of our small theory group are Subrahmanyan Chandrasekhar and Eugene Parker.'

People often tell me that I must have been extremely lucky to have worked in such an environment. I usually completely disagree. When you are a student struggling with your PhD thesis which seems to be going nowhere, it is very useful to have around you senior students or postdoctoral fellows whom you can think of as your role models and on whose shoulders you can cry when the research problem you are trying does not work out. As long as Tom Bogdan was there, he often provided me with moral support. After he left, there was no one like that. I could not possibly think of Chandrasekhar or Parker as my role models. At one point, I was fairly close to a nervous breakdown. It was a miracle that I did not have one.

I found Parker to be an extraordinary human being in addition to being an extraordinary scientist. He had a tremendous sympathy for the underdogs of the world everywhere. His face would turn red with anger when he denounced the terrible things which Ronald Reagan's government was doing in Central America. Although he avoided too much travelling, he never missed opportunities to visit countries like India, China and the Soviet Union where people were doing science under difficult conditions. He was always a keen observer of science done in countries like India and China, and wanted these countries to flourish (they were economically impoverished countries in the early 1980s). While my admiration for Parker kept increasing, my research did not progress!

Parker told me in the very beginning that my PhD thesis was mine and I had to figure out what I wanted to do. I could come and discuss with him at any time. But he would not tell me what to do. Parker never touched a computer. All his calculations were done with paper and pencil. It was expected that I too would get initiated into this school of research. I had no idea how one chose research problems. I read the recent literature of the field quite extensively and tried to do three calculations one after another, taking up several months. None of these yielded any worthwhile results. When Parker told me that I could discuss with him at any time, he really meant it. Whenever I wanted to discuss with Parker throughout the four years I worked with him, if he did not have a visitor in his room, he would always set aside what he was doing and discuss with me. I do not recall his telling me even once: 'Today I am busy, can we meet some other time?' At that young age, I did not fully realize that this was very unusual for an academic of Parker's stature. I feel amazed when I think about this now. Parker always gave you the impression that he had all the time in the world to talk to you and was never in a hurry to get back to some urgent work after you left.

My scientific discussions with Parker did not help me in any way. Most of us need some mental preparation to do a scientific calculation and cannot start doing a calculation all of a sudden. Parker had the unbelievable ability to concentrate all his intellectual powers at a moment's notice. Soon after I would start telling him the calculations I was doing, he would spring to the blackboard and say: 'Let me see how this can be done. Let us start from the beginning.' Then he would formulate the problem in his own way and start doing calculations on the blackboard, often recalling from memory some complicated formulae involving mathematical functions such as Bessel functions or Legendre polynomials. Sometimes he would reproduce in half an hour what I had taken weeks to do and would then proceed further to do things which I could not do. I could only gaze at him in amazement and awe. I did not know that Parker was probably the *only* theoretical solar physicist in the entire world who could do such calculations on the blackboard at such speed. I thought that one had to be able to do such calculations on the blackboard to be a scientist in this field and I knew that I could not do such things. I would return from each such scientific session totally devastated and demoralized.

After about one-and-a-half years, I was convinced that I did not have the capabilities to become a theoretical astrophysicist. Several of my classmates who worked with other professors already had their first papers published. I always had a deep interest in the philosophy of science. The University of Chicago had very eminent professors in this field. I had attended some of their courses and interacted with them quite extensively. I had almost completed arrangements for making a transfer from theoretical astrophysics to philosophy of science. When I went to tell Parker about my plan, he probably understood what was going through my mind. He asked me to sit down and then told me in an unusually gentle tone: 'If you want to make this transfer for the positive reason that you find philosophy of science much more appealing than theoretical astrophysics, then you will have my best wishes. But please do not make the transfer for the negative reason that you are incapable of doing astrophysics.' Then he told me that he had been closely watching how I was doing my calculations. Although nothing had worked so far, he said that he was still very impressed with me and he expected that something would work out soon. Finally he told me something remarkable: 'You should think carefully what you expect out of your life. If your wish is to become the second Subrahmanyan Chandrasekhar, then I have to tell you that this is not going to happen. But I see no reason why you will not become a top theoretical solar physicist of the world if you work with dedication.' After this conversation with Parker, I decided to give my calculations a last try. I made up my mind that, if the next calculation did not work, then I would quit astrophysics. Well, my next calculation finally worked and I got my first paper in *The Astrophysical Journal*. I became an astrophysicist instead of a philosopher of science by a whisker.

Now I shall mention one more thing about Parker. Although scientific collaborations and multi-author papers were becoming the norm of scientific research by the end of Parker's active career, more than 90% of his papers are single-author papers. One often hears complaints about some senior persons who want to put their names in papers by younger colleagues without contributing anything significant. Parker was the exact opposite. Many younger persons working in Parker's group would often have liked to include him as co-author because they would then be able to claim in later life that they have written papers with Parker. But Parker told me that he never agreed to become a co-author in a paper unless he had done all the calculations in the paper once by himself! To my eternal regret, I never managed to have a joint paper with him.

5.5 Flux Rings and the Coriolis Force

Anybody familiar with the physics academic world would know that it is customary for a young academic to spend two or three years in a temporary postdoctoral position after PhD before establishing himself/herself and looking for a more permanent job. Many good research groups around the world would have openings for such postdoctoral positions. After completing my PhD in Chicago, I took up a 2-year post-doctoral position at the High Altitude Observatory (HAO) in Boulder, which had one of the most vibrant solar physics research groups in the USA. It was at HAO that John Eddy about a decade earlier had initiated his famous work on the Maunder minimum (Section 2.2), but he was thrown out of his job in the middle of this work! Although I realized that numerical simulations were going to be increasingly important in theoretical solar physics research, I did not have any opportunity of learning numerical simulations when I was in Chicago. So I thought that the most profitable way of spending my two years in HAO would be to learn that. Often it is said that doing numerical simulations is as much an art as a science. Learning numerical simulations completely on one's own is close to an impossibility. So I wanted to work with somebody who had done numerical simulations before and who could teach me the tricks of the trade.

Peter Gilman was someone in HAO who had published some impressive numerical simulations of convection and dynamo process inside the sun. I asked him if I could work with him and he agreed enthusiastically. It was then that I discovered to my astonishment that Gilman had probably not even looked at a computer code for years, but would get other (usually younger) persons to do the work for him. This was my first encounter with such a person, who was so completely different from Gene Parker. I quickly realized that I could not possibly learn how to do numerical simulations from Gilman himself. However, there was an amazing computer programmer in HAO named Jack Miller. He did not have any formal training in physics research. But he could write the numerical code if a scientist just gave him the differential equations to be solved. I came to know that Jack was the person who wrote the computer code for Gilman's dynamo work. Jack and his wife Barbara were very interested in Indian cuisine. I became quite close to them, though I could not teach them any Indian cooking. Barbara could already cook some Indian dishes which were beyond my limited culinary skills. It was from Jack that I learnt how to do numerical simulations.

Gilman told me that he was bothered by one scientific question and I might find it worthwhile to look into it. If you look carefully at the butterfly diagram in Figure 2.3, you will see that sunspots are usually not found very close to the equator. There is a narrow band near the equator where sunspots are rarely found. One possible reason for this could be that toroidal magnetic flux tubes at the bottom of the convection zone from which sunspots arise may not form very near the equator. The other possibility is that toroidal magnetic flux tubes at the bottom of the convection zone form near the equator, but parts of these flux tubes rising through the convection zone to form sunspots may get deflected by some force to move away from the equator so that sunspots do not appear very close to the equator. Gilman felt that it would be interesting to look at the second possibility. What force could have deflected a rising flux tube? Physicists have been aware since the beginning of the nineteenth century of a force called the Coriolis force (named after Gaspard-Gustave Coriolis who discussed this force around 1835), which could have been the culprit. This force arises when something moves on a rotating body. For example, the earth rotates about its axis in 24 hours. So moving objects on the surface of the earth are subject to the Coriolis force. If the motion lasts much less than 24 hours, then usually the effect of the Coriolis force is not significant. However, things moving freely (i.e. not constrained to move in particular paths) over large distances for times longer than 24 hours are affected by the Coriolis force. This force has profound effects on the large-scale oceanic and atmospheric circulations. As the equatorial region of the earth gets heated by the sun's rays, the hot air there rises up in the atmosphere. As a result, colder air from more temperate latitudes blows towards the equator. One would expect the wind in the northern hemisphere to blow towards the south and the wind in the southern hemisphere to blow towards the north. It is found, however, that the wind in the northern hemisphere blows towards the south-west and the wind in the southern hemisphere blows towards the north-west. It is the Coriolis force due to the earth's rotation which causes this deflection of the wind system. Gilman and I wondered whether the Coriolis force of the sun would have any effect on the rising flux tubes.

Since the sun is rotating about its axis, there is no doubt that this rotation can give rise to the Coriolis force that would act on something moving in the sun. Until that time, nobody had looked into the question of whether this Coriolis force would have any effect on flux tubes rising through the convection zone. This was completely uncharted territory, which I agreed to explore. Neither Gilman nor I at that time had the slightest suspicion that this could turn out to be an investigation which would eventually shake the foundations of theoretical solar physics.

Anybody who has ever done any mathematical physics calculation knows that calculations for symmetric situations are much easier than calculations for non-symmetric situations. The magnetic field configuration shown in Figure 2.8(a) is certainly a non-symmetric configuration with one part of the magnetic flux ring coming out. Suppose the whole magnetic flux ring was lying just above the bottom of the convection zone at an instant of time. Then magnetic buoyancy would be equally strong in all parts of the flux ring (unlike what happens in Figure 5.3) and the whole flux ring will rise symmetrically through the convection zone. This is a situation for which calculations based on MHD equations are much simpler. To understand the effect of the Coriolis force, I decided to first study this symmetric situation. With some tips from Jack Miller, I managed to write a computer code for this problem. To run the code, I had to specify some initial value of the magnetic field inside the flux ring. All solar dynamo theorists at that time were convinced that the magnetic field in the sun's interior could not be stronger than about 1 tesla. The reason behind this conclusion will be discussed in the next chapter. I started running my computer code by assigning a value of 1 tesla to the initial magnetic field inside the flux ring at the bottom of the convection zone.

I expected the Coriolis force to deflect the rising flux ring in such a way that a point of the ring would follow a trajectory shown by the dashed line in Figure 5.5. Then we would be able to explain why sunspots do not appear very close to the equator. Most computer terminals in those days were rather primitive. My terminal did not have any graphics display facility. I could see a plot only after the output of my plotting programme was printed using a printer that was kept in

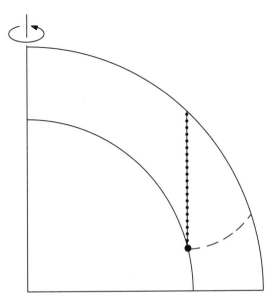

Figure 5.5 The trajectories of magnetic flux rings rising through the sun's convection zone starting from its bottom. The line with dots shows the trajectory for an initial magnetic field of 1 tesla inside the flux ring, whereas the dashed line shows the trajectory for an initial magnetic field of 10 tesla inside the flux ring.

a corner of HAO far away from my office. After my code made its first run and I gave the command to print the first plot, I almost ran to the printer in eager anticipation. I was greatly disappointed to see that the trajectory of the rising flux ring seemed like a straight line parallel to the sun's rotation axis, as shown by the line with dots in Figure 5.5. My first reaction was that I had made a mistake in the plotting programme. However, after checking it thoroughly, I could not find any mistake. I was then convinced that the mistake was in the code, which would be harder to detect. I spent the next few days checking the code and still could not find anything wrong. Then a strange thought came to my mind: could the plot be correct! I decided to run the code for a much stronger initial magnetic field of 10 tesla. Lo and behold, the trajectory now turned out to be exactly like the dashed line in Figure 5.5.

I was completely baffled. Sunspots always appear at latitudes lower than 40° (see the butterfly diagram in Figure 2.3). But the dotted trajectory for 1 tesla shown in Figure 5.5 reaches the sun's surface at a much higher latitude. If the solar dynamo could not produce magnetic fields stronger than 1 tesla, as emphatically asserted by all dynamo theorists at that time, and if the effect of the Coriolis force was as strong as our calculations suggested, then we should see sunspots only at rather high latitudes! In order to produce sunspots at low latitudes where they are seen, the magnetic field in the interior of the sun would have to be as strong as 10 tesla—which all dynamo theorists at that time considered to be an impossibility. If the Coriolis force was really as strong as we had found, then all the theoretical models of the solar dynamo available at that time (including some of my own papers) were most probably completely wrong.

Neither Gilman nor I knew what to make of these results, but we had a hunch that we had accidentally stumbled upon something important and decided to present the results in a paper. The Choudhuri and Gilman 1987 paper appeared in *The Astrophysical Journal*.[4] To our great disappointment, it did not even receive the slightest attention in the solar physics community. After I had become more experienced a few years later, I realized that most scientists easily get excited when they are shown some results which they have been expecting. When confronted suddenly with something which no one expected, people often do not know how to respond. I did not understand this at that young age and was bothered why others did not seem worried at all about the strange results we had found. I got an opportunity to present these results at a meeting of the American Astronomical Society, and hoped that, if people heard me, they might realize that something was seriously amiss somewhere. I like to flatter myself with the thought that I am not usually the worst speaker at a scientific meeting. But, when I was mid-way through my talk, I looked around and almost every member of the audience appeared to be thoroughly bored and uninterested in what I was saying. As I struggled to finish my talk, my eyes suddenly fell on one member of the audience—Robert (Bob) Rosner, who was in the faculty of Harvard University at that time. His face seemed to glow with excitement and he was absorbing every word I said. Bob came over to me during the tea break after the session and said: 'It was an excellent talk. But I do not believe what you presented. You must have overlooked something in your calculations. If your results are correct, that would require a major paradigm shift in theoretical solar physics.' Manfred Schüssler is a German scientist who, a few years later, did a lot of important work on the effect of the Coriolis force on rising flux tubes.

I was once telling him about the poor initial response to our paper—the first work on this subject. Manfred told me that he had read our paper immediately after its publication and was very disturbed. In fact, he said that he had not been so disturbed by a scientific paper in many years. But he did not know what to do with our results. It took him a few years to come to terms with these results and only then did he start working on this topic.

My own view on this subject at that time was exactly similar to the view Bob Rosner expressed after my talk. The arguments given by dynamo theorists at that time that the magnetic field inside the sun could not be stronger than 1 tesla seemed very compelling to me. So, there must have been something missing in my calculations. If this missing something was included, then presumably the effect of the Coriolis force would not be so strong and flux rings with magnetic field of 1 tesla would emerge at low latitudes. But what could this missing something be? I had done the calculations for a symmetric flux ring. In reality, however, we know that only a part of the flux ring rises up through the convection zone, as shown in Figure 2.8(a). Was it possible that the effect of the Coriolis force would be much less if we considered a small part of the flux ring rising, rather than the whole symmetric flux ring? I knew that the next step would be to do a simulation of a part of the flux ring rising and to study the effect of the Coriolis force on it. Doing such a calculation starting from the basic equations of MHD would be extremely difficult. I thought about this problem for several weeks, without being able to make a clear formulation of the problem or without being able to come up with a feasible scheme for the numerical simulation.

Kazunari Shibata, a Japanese scientist slightly older than me, stopped at HAO for a short visit. Later Shibata turned out to be probably Asia's most influential theoretical solar physicist of our generation and trained several outstanding students. That was Shibata's first trip abroad and he spoke English haltingly. But still we had a very long scientific discussion. I told Shibata where I had got stuck and that I did not know how to proceed. Shibata told me that he had recently seen a 'preprint' on the subject, but could not remember the author's name. Perhaps I need to explain what a 'preprint' was, because even young scientists nowadays may not have heard of preprints. There are many electronic archives nowadays where one can put a scientific paper as soon as it is written and then anybody can see it. I am talking about a time before the internet. In those days, many scientists would send copies of their manuscripts (before they appeared in journals) to research groups where people would be interested in their work. Many research groups used to have preprint racks where preprints received in the last few months would be kept. Shibata suggested that we could go to the preprint rack of HAO and then he could look for that preprint there. We went through the several piles of preprints kept in the preprint rack of HAO and came to the last pile. Suddenly Shibata exclaimed joyously and handed me the preprint we had been looking for. The author had the exotic Spanish name Fernando Moreno-Insertis. As soon as I glanced through the pages, I knew that this preprint would give me the clue to solve my problem. I almost rushed back to my room with the preprint, without even adequately thanking Shibata. I wonder if Shibata considered me a rude and ill-mannered person. He took so much trouble finding that preprint and I did not even want to talk to him any more once it was found.

What Fernando Moreno-Insertis had done was basically the first comprehensive numerical simulation of the process shown in Figure 5.3.[5] If a part of a flux tube came out into the convection zone, Fernando's simulation showed how that part would rise through the convection zone. Although everybody expected that the magnetic configuration shown in Figure 5.3(a) would after some time evolve into the magnetic configuration shown in Figure 5.3(b), nobody had done a detailed calculation of this process before Fernando. In Fernando's simulation, the flux tube was always assumed to lie in a plane and the effect of the Coriolis force was not included. All I now had to do was to follow Fernando's method to do a simulation of the magnetic configuration shown in Figure 2.8(a) by including the Coriolis force. Doing such a simulation starting from the basic MHD equations is very complicated (this has been done by Yuhong Fan and others in the last few years). Hendrik Spruit, a Dutch scientist who worked in Munich, had derived an equation for flux tubes from the basic MHD equations. Fernando devised a scheme for solving Spruit's equation numerically, which was much easier than solving MHD equations. I read Fernando's paper repeatedly to understand the method and then started writing my computer code to do the first detailed numerical simulation of the magnetic configuration shown in Figure 2.8(a).

5.6 Back to the Homeland

My 2-year job in HAO was coming to an end and I had to decide what to do next. Indian academics in my age group who had done our PhDs in the USA faced the Hamletish question of whether to return to India or not. One often hears heated arguments whether a scientist has a moral obligation to work in his or her own country to help that country's scientific development. This is a complex ethical question to which I shall not venture to provide a general answer. I can only talk about my personal decision. I was born a few years after Indian independence. Mahatma Gandhi was my childhood hero and I grew up with the idealism that we have to build our newly independent nation. Then corruption was not as endemic in the Indian political life as it has become now. We had an enormous respect for our first two Prime Ministers—Jawaharlal Nehru and Lal Bahadur Shastri—which educated Indians never had for any of their successors. In spite of India's terrible poverty, I was proud of India's thriving democratic tradition and that was one of the main reasons behind my decision to return to India. Had I been from a country under military dictatorship or under the sway of religious fundamentalism, I sometimes wonder what decision I would have taken.

When I was working in the USA, I was constantly bothered by the question whether I would be able to get a suitable job in India (academic jobs were few and far between in India in those days) that would enable me to carry on my research. I was overjoyed when I received the offer of a lecturer's position in the newly-formed astrophysics group at the Indian Institute of Science (IISc) in Bangalore. On the day I was going to post my acceptance letter (those were the days before e-mail), I met Boon Chye Low, a distinguished theoretical solar physicist working in HAO, in the corridor and told him about my decision. Boon Chye is from Singapore and did his PhD with Gene Parker a few years before me. Boon Chye took me to his office, told me various things about his experience with Singapore and then said strongly: 'Things are going well with you in USA. Surely Gene Parker will help you in getting a good job in USA. We shall all help you. You should give up this crazy idea of going back to India at less than one-tenth of the salary you are getting here. If you go back to India, that will be the end of your scientific career.' Since I have always looked up to Boon Chye like my elder brother and I knew that he was my genuine well-wisher, I was very disturbed by this conversation. I returned to my office and started thinking what I should do. I then decided to telephone Gene Parker and told him about my conversation with Boon Chye. Gene listened to me patiently and then asked: 'Do you really—I mean *really*—want to return to India?' I said that I had the dream of building up a research group in India, but I was totally unsure if I had the intellectual capability of thinking up good research problems completely on my own in an isolated place. I was only 30 and had limited research experience. After I had explained everything, there was a silence on the telephone line. I wondered if Gene was still there at the other end. Then I heard him utter three short sentences slowly and clearly—three sentences which determined the course of my life and which still ring in my ears: 'Arnab, we each have only one life to live. Go, chase your dream. I have confidence in you.' I thanked Gene, put down the telephone and posted my acceptance letter.

Barely a month before I was to leave for India, my computer code for studying the evolution of the magnetic configuration of Figure 2.8(a) became ready. I was very busy packing and could only make a few trial runs. I found that, even when only a part of the flux ring was allowed to rise, the effect of the Coriolis force remained equally strong. If the initial magnetic field inside the flux ring was assumed to be 1 tesla, the rising part of the ring followed the trajectory shown by the dotted line in Figure 5.5. I had to make the initial magnetic field as strong as 10 tesla if I wanted magnetic buoyancy to be strong enough to overcome the Coriolis force, making the top of the flux ring follow the dashed trajectory in Figure 5.5. But I had no time to investigate the problem in detail. I explained the code to Jack Miller. We made the plan that I would send Jack any instructions for running the code (by air mail, since e-mail networks still had not come into being) and Jack would then send me back the results. Hoping that I also would be able to run the code in Bangalore, Jack copied the code in a large magnetic tape of about one foot diameter. The floppy disc had not yet started being widely used and we could not even imagine that in a few years it would be possible to send such documents simply through e-mail. We hoped that I would be able to mount the magnetic tape in some computer at the Indian Institute of Science (IISc) and retrieve the code. Nowadays a code like this can easily run on a PC, but in the late 1980s we needed the most sophisticated computers of that time to run it.

I returned to India in 1987 about three or four years before India's remarkable economic growth started. The Indian economy had been stagnant for many decades and the cost of a PC in India (after adding the stiff import duty) was more than one year's salary for an entry-level academic like me. The science funding situation was also quite dismal. It was no wonder that it was extremely difficult even to do modest computations in IISc at that time. The only plan I could make was that, after settling down, I would send a detailed list to Jack Miller about all the runs he should do with my code in HAO.

Then came a bolt from the blue. I suddenly got a letter from Jack saying that his job in HAO had been terminated and that he was shifting out of Boulder in search of another job. I was stupefied. Everybody in HAO knew that Jack was extraordinarily competent at his job and some of the senior scientists there had their reputations built on the basis of the codes Jack had written. It seemed to me like extreme ingratitude and meanness on the part of an organization to fire a person like this (Peter Gilman was one of the top administrative bosses in HAO at that time). There was no hope of running my code in IISc. Now I knew that there was no possibility of running the code in HAO also. Nearly one year I had spent on developing the code simply seemed to have gone down the drain. I was terribly depressed.

As I mentioned, I had finished developing the code about a month before I left HAO and I had about ten plots based on the initial trial runs. These plots clearly showed that the Coriolis force was still very strong when we considered only a part of the flux ring rising. Out of desperation, I decided to write a paper based on these plots, because I knew that there was no chance of running my code again in the near future. Normally, after developing a code, I would test it thoroughly and make many runs giving different values of the important parameters. Only after gaining an insight into the problem on the basis of many runs, would I eventually write a paper presenting the best results. This was no longer possible. The main worry was the refereeing process. When a paper is submitted to a good journal, the editor of the journal always takes the opinion of another scientist in the field (called a 'referee') whether the paper is worth publishing. If the referee of my paper asked for some clarifications which would require running the code again (a very normal thing for a referee to do), then I would be in deep trouble. Most of my papers had so far appeared in *The Astrophysical Journal*. I decided to submit this paper to the journal *Solar Physics*. The editors of this journal were eminent solar physicists and I thought that they would be more considerate with me if the need arose to explain my situation in response to the referee's criticisms. Luckily the paper went to a very generous referee who recommended its publication as it was. Thus appeared an account of the first numerical simulation of the magnetic configuration depicted in Figure 2.8(a).[6] The plots given in that paper were the only plots obtained with my code that I had for many years.

5.7 Joy's Law Explained

The research facilities and conditions in IISc during my initial years were so inadequate for the kind of research I wanted to do that I was sure for a long time that Boon Chye's prediction about the end of my scientific career would come true. The only redeeming feature was that I was in a physics department with extremely brilliant colleagues and students. I also found that I enjoyed teaching very much and developed a good rapport with the students in our department (many of whom are very eminent academics today). But keeping abreast with what was happening in our field was extremely difficult. In those pre-internet days, one had to depend on the library for the latest publications. The library funds were limited and even the journal *Solar Physics*—one of the most important journals in our field—was not kept in our library. Another way of knowing what was happening in one's field was to attend international conferences. The cost of attending one conference outside India in those days was comparable to my one year's salary. I simply had no funds for attending international conferences. A few years after I joined IISc, Nigel Weiss organized an important conference on sunspots in Cambridge and asked me to give an invited talk on an aspect of sunspots on which I had done research. Nigel knew of my difficult situation and managed to find full funds for my trip to Cambridge. Almost all the important scientists working on sunspots were present at that conference. The talk I gave in Cambridge was my only talk at an international conference outside India during my first 12 years in IISc.

After my first PhD student Sydney D'Silva started working with me, things looked a little brighter. Sydney had a Portuguese last name because his forefathers living in the Malabar coast of India were converted to Christianity by Portuguese missionaries, though there was no actual Portuguese blood in the family. I still believed that the effect of the Coriolis force in my simulations was so strong because something had been overlooked. I asked Sydney to look for this missing something. With the modest computational facilities we had, we could only do the simulation of a symmetric flux ring (as was done in the Choudhuri and Gilman 1987 paper) which required much less computational resources compared to what was required for a simulation in which a part of the flux ring only was buoyant. We could show that some effects could suppress the Coriolis force if they were sufficiently strong. But we were not sure if these other effects could be so strong inside the sun and the results were inconclusive.

With the passage of every year, the computational facilities in IISc kept improving and Sydney told me that we should make an effort to run my big code which I had developed before leaving HAO and which had not been run since then. I showed Sydney the magnetic tape of about one foot diameter lying on a bookshelf idly for many years in which my code had once been copied. I wondered if the tape would still be usable or if it would have been by now damaged by the various species of fungus which flourished in the Indian weather conditions. In those days, tape drives in computers were not very standardized. Sydney made trips to the computer centres of different departments of IISc and then informed me that one computer in a department had a tape drive which could probably read my tape. We walked over to that department with the tape. The computer operator there was a very friendly young guy. He managed to mount the tape on the drive. We gave a command to list the files on the tape, wondering if we would simply get an error message that the tape was unreadable. I missed a heartbeat when the names of all the files appeared on the screen, including the filename of my code. With trembling fingers, I typed the command for opening the file containing my code. Then, after a gap of nearly five years, I again saw my code on a computer monitor! I felt that I was looking at my long-lost child. Suddenly I realized that tears were welling up in my eyes. Since I did not want Sydney or the computer operator to see my tears, I pretended that I had to use the toilet and rushed away.

Now our job was to find a computer on which my code could run. The condensed matter theory group in our department had a Vax computer and I asked them if we could use it. In those days, computers were not inter-linked. To use a computer, one had to physically sit in front of one of the four or five terminals kept next to it in the 'computer room'. The condensed matter colleagues were very sympathetic, but they also had limited facilities and their terminals were always occupied by their students during office hours. They told me that my students could use their computer after 10 pm at night. Luckily Sydney was a night bird. He would come to work in the computer room after 10 pm, work till the wee hours in the morning and go to bed when light appeared in the eastern sky. I rarely saw him during office hours.

I have mentioned Joy's law in Section 2.5 when discussing the magnetogram map shown in Figure 2.7. The lines joining the centres of sunspot pairs with opposite polarity seem nearly parallel to the sun's equator, but not exactly parallel. The sunspot on the right side in the pair appears slightly closer to the equator. These tilts of sunspot pairs increase with increasing latitudes. This law that tilts increase with latitude was discovered by Alfred Joy in 1919. Although it was an extremely important observational law in the study of sunspots, so far it had no proper theoretical explanation. Some solar physicists conjectured that the action of the Coriolis force on the rising flux tubes could have caused these tilts. But, apart from giving hand-waving arguments, nobody had done any detailed calculations till that time. Now that we had a numerical code for studying the effects of the Coriolis force on a part of the flux ring rising due to magnetic buoyancy, Sydney and I decided to check whether our simulations would throw any light on Joy's law. I still had a strong faith in the arguments given by solar dynamo theorists that magnetic fields in the interior of the sun could be at most 1 tesla. According to my first simulations, parts of flux rings with such magnetic fields would rise parallel to the rotation axis and emerge at high latitudes. I told Sydney that he should explore if the Coriolis force could be suppressed enough to make the parts of flux rings with 1-tesla magnetic fields appear in lower latitudes, but there should be some effect of the Coriolis force to explain Joy's law. To my dismay and exasperation, Sydney suddenly decided to do the calculations by assuming the magnetic field in the sun's interior to be 10 tesla, since only such magnetic fields would produce sunspots at the correct latitudes according to my previous simulations. I told Sydney: 'If you want to run the code with 10-tesla magnetic fields, that is your wish. But you will be wasting precious computer time. I do not think that it is possible for the solar dynamo to produce magnetic fields stronger than 1 tesla in the sun's interior.' Luckily Sydney decided to ignore my advice!

After a few days, Sydney came to show me some plots he had obtained. My eyes almost popped out of my head as I looked at them. I at once knew that at last we had before us the first theoretical explanation

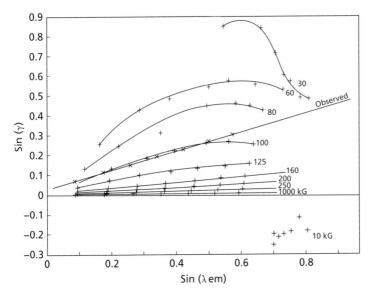

Figure 5.6 A figure from the D'Silva and Choudhuri 1993 paper showing how the tilts γ of bipolar sunspot pairs vary with emergence latitude λ_{em}. The straight line is the best fit to the observational data of Joy's law. The curves were obtained from numerical simulations assuming different values of the initial magnetic field inside the flux ring (indicated in kilogauss).

of Joy's law discovered three-quarters of a century earlier. Figure 5.6 shows the final combined version of the plots which Sydney showed me that day. He had made many runs on the computer assuming different initial values of the magnetic field inside the magnetic flux ring. For a particular value of the initial magnetic field, we would consider flux rings at different latitudes at the bottom of the convection zone and would allow parts of these flux rings to rise due to magnetic buoyancy. They would reach the surface at different latitudes with different tilts. For a particular value of the initial magnetic field, one would find a particular value of the tilt at a certain emergence latitude on the sun's surface. The different curves in Figure 5.6 are the plots of tilt versus latitude obtained for different values of the initial magnetic field. The values of the initial magnetic field are indicated next to each curve in kilogauss, which is 0.1 tesla. The straight line is the best fit to the observational data representing Joy's law. You can see that the theoretical curve for 100 kilogauss or 10 tesla is very close to this straight line representing observational data. If the initial magnetic field inside flux rings is very strong (1000 kilogauss or 100 tesla), then we see in Figure 5.6 that the tilts are very small because the strong magnetic buoyancy of such a strong magnetic field would completely overpower the Coriolis force, which would not be able to produce significant tilts. On the other hand, if the initial magnetic field is very weak (such as 30 kilogauss or 3 tesla), then the Coriolis force makes the rising part of the flux ring emerge at high latitudes so that the plots for such initial magnetic fields in Figure 5.6 do not extend to low latitudes and do not match the observations at all. Only when the initial magnetic field inside the magnetic flux ring lies in a narrow range around 10 tesla, did we find that the Coriolis force is just strong enough to produce the correct tilts, but not strong enough to divert the rising parts of flux rings away from low latitudes.

On the day that Sydney showed me the plots which were eventually subsumed in Figure 5.6, I was at last convinced that there had to be something seriously wrong with the dynamo arguments that the magnetic field in the sun's interior is not stronger than 1 tesla. The successful explanation of Joy's law now made the conclusion virtually inescapable that the magnetic field in the sun's interior has to be closer to 10 tesla. We wrote up the D'Silva and Choudhuri 1993 paper[7] and wondered how it would be received by the solar physics community. Unlike what happened with the Choudhuri and Gilman 1987 paper, we found that the D'Silva and Choudhuri 1993 paper was very soon hailed as a landmark by many solar physicists. For many years, solar physicists had been waiting for a theoretical explanation of Joy's law and our paper provided this explanation—unlike the Choudhuri and Gilman 1987 paper which presented results for which nobody was prepared at that time.

Two other groups in two different continents also started doing similar calculations at that time. The senior persons in the European group were Fernando Moreno-Insertis and Manfred Schüssler, whereas the senior persons in the American group were Yuhong Fan and George Fisher. They confirmed the results of the D'Silva and Choudhuri 1993 paper and explored many other interesting aspects of the problem. In these days of exploding literature, often publishing a paper in a good journal is not enough. One has to go to scientific conferences to talk about one's work so that people come to know what one is doing.

As I already mentioned, I had no funds at that time to attend international conferences. However, I kept on hearing from various people that Moreno-Insertis, Schüssler, Fan and Fisher were giving talks in important international conferences highlighting our work. They were our rivals in research, but our work became widely known because of their generous presentations of it. Nothing can be more wonderful and more touching for a scientist working under difficult conditions than to have his rivals as his brand ambassadors! The personal relationships of the three groups working in this field at about the same time were always very cordial. When I was working on the solar dynamo problem a few years later, I fondly and nostalgically used to recall the days when our rival groups were making our work on magnetic buoyancy widely known. The relationships amongst some of the groups working on the solar dynamo a few years later were often acrimonious.

5.8 Quo Vadis?

The magnetic buoyancy calculations of our three groups finally convinced the solar physics community at large that the magnetic field inside the sun has to be much stronger than what the dynamo theorists at that time believed to be the case. It became clear that the solar dynamo models popular at that time had to be discarded. In a few years, there emerged a new kind of dynamo model (although its roots go back to the 1960s) that could account for the strong magnetic field in the sun's interior. We shall discuss these developments in the next two chapters. You can now check the central dogma that I had put forth in Section 2.5. You will realize that we have already explained a substantial part of this central dogma. We have seen how the toroidal magnetic field may arise from the poloidal magnetic field and then how we get sunspots forming from this toroidal magnetic field. Now our job is to explain the remaining part of the central dogma. We have to explain how the poloidal magnetic field arises out of the toroidal magnetic field, leading to the 11-year cycle of oscillations between the two components of the magnetic field. This is the heart of solar dynamo theory. Chapter 6 will describe how the theory evolved and why the initial models suggested that the magnetic field inside the sun could not be stronger than 1 tesla. Then in Chapter 7 we shall discuss the more modern models of the solar dynamo that try to account for the much stronger magnetic field inside the sun's interior.

Since I have written a lot about the difficulties I faced in my early years in India, let me end with a few comments on how the situation evolved in India. The economic condition of India started improving rapidly from the early 1990s after having remained stagnant for a long time. This economic boom initially had an almost adverse effect on scientific research. After incomes in many other professions started increasing rapidly, academic salaries remained virtually static for a few years and young people who wanted a good standard of living almost stopped coming to academics. A turnaround started in about 2000 when academic salaries finally started rising and the science funding situation in India improved significantly. A junior faculty member in a good place in India can now have most of the facilities and opportunities which his or her counterparts in other countries will have.

6

Dynamos in the Sky

6.1 Breaking the Symmetry with Turbulence

This is a good time to take another look at the central dogma enunciated in Section 2.5, and then to take stock of where we are. I have already explained a part of the central dogma. If there is any poloidal magnetic field inside the sun, Figure 4.11 shows how Alfvén's theorem of flux freezing leads to the creation of a toroidal magnetic field. I have discussed in the previous chapter how the ideas of magnetoconvection and magnetic buoyancy can be used to explain the formation of toroidal flux tubes and then to explain how parts of these flux tubes may rise to produce bipolar sunspot pairs. We have thus seen that theoretical explanations can be provided for many aspects of sunspots if we begin with a poloidal magnetic field as shown in Figure 4.11. The next question before us is: where does this poloidal magnetic field come from? Unless there is a mechanism for producing the poloidal magnetic field, even if we somehow had a poloidal magnetic field to begin with, it would decay away because the currents which must be flowing through the solar plasma to create this poloidal magnetic field would decay away due to the resistance in the plasma. The oscillation between the toroidal and poloidal magnetic fields, as seen in Figure 2.11, strongly suggests that the poloidal magnetic field is produced from the toroidal magnetic field. Now we come to the crucial question of how the poloidal magnetic field is produced from the toroidal magnetic field and how the oscillation between these two types of fields arises. Once we answer these questions in this and the next chapter, you will have a complete explanation of the central dogma of solar dynamo theory. This chapter will introduce some of the classical ideas of this field. According to these classical ideas, the toroidal magnetic field inside the sun cannot be stronger than about 1 tesla. As I discussed in the previous chapter, magnetic buoyancy calculations done by us and other groups suggested that the toroidal magnetic field in the interior of the sun should actually be

Figure 6.1 Thomas George Cowling (1906–1990), whose anti-dynamo theorem was a landmark in the historical development of dynamo theory.

closer to 10 tesla. Chapter 7 will describe attempts to develop a new kind of solar dynamo model that can account for this stronger toroidal magnetic field.

I have written a little in Section 1.4 about early efforts towards building models of the solar dynamo. I mentioned that a landmark in the historical development of dynamo theory was Cowling's anti-dynamo theorem discovered in 1933. Sufficiently symmetric plasma motions cannot sustain a sufficiently symmetric magnetic field. If you take another look at Figure 4.11, you can convince yourself that generation of the toroidal magnetic field from the poloidal magnetic field can be a fairly symmetric process. If we start with a symmetric poloidal magnetic field as shown in Figure 2.10(b), differential rotation can stretch it out to generate a symmetric toroidal magnetic field like that shown in Figure 2.10(a). This means that the complementary process of generation of the poloidal magnetic field from the toroidal magnetic field cannot be symmetric and has to be much more complex. If this process were also symmetric, that would have clearly violated Cowling's theorem. Now, within the sun's convection zone, we have hot gases going up in some places and cold gases coming down in other places. If we were to fix our attention on a point inside the sun's convection zone, then we would find that the nature of gas motion there would keep changing randomly with time. Sometimes we would have hot gas going up there and sometimes cold gas coming down. This kind of situation inside a fluid in which the nature of fluid motions at points inside the fluid keeps changing randomly is called *turbulence*. Certainly the turbulence inside the sun's convection zone involves non-symmetric motions of the gas. Parker realized in his famous 1955 paper on the dynamo that one can invoke turbulence to break the symmetry and get around Cowling's theorem.

Let us begin with some discussion on turbulence. A systematic study of turbulence began in 1883 when Osborne Reynolds carried out a rather simple experiment.[1] He made water flow through a horizontal pipe and injected a dye at a point in the water just before it entered the horizontal pipe. If the water was not flowing too fast, Reynolds found that the dye was just carried downstream in a straight line. This kind of a regular flow of water is called *laminar flow*. However, when Reynolds made the water flow faster by applying a larger pressure, something interesting happened. The dye started spreading inside the pipe as it was carried downstream. Figure 6.2 shows Reynolds's original sketch taken from his paper showing how the dye was carried downstream in the two situations of laminar flow and turbulent flow. The spreading of the

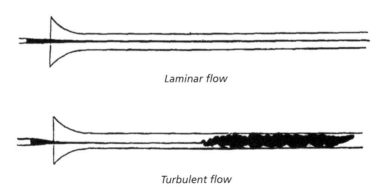

Laminar flow

Turbulent flow

Figure 6.2 Reynolds's sketches showing what happens to a dye injected into water that flows through a horizontal pipe.

dye on making water flow faster indicated that the flow inside the pipe had become irregular and that turbulence had set in.

Let us now go more than a century back in time before Reynolds's experiment. At that time the Swiss mathematician Leonhard Euler, one of the most illustrious names in the history of mathematics, began formulating the mathematical equation for studying fluid flows theoretically. Euler lived at a time when some of the enlightened rulers of Europe thought that keeping a philosopher or a mathematician in the royal court would enhance the prestige of the court. King Frederick the Great of Prussia, who was the patron of the philosopher Voltaire, was one such enlightened ruler. Euler was employed in Frederick's court. When Frederick wanted to have some marvellous water fountains in the garden of the Sanssouci Palace in his capital Potsdam, Euler started investigating the flow of water through fountains. Euler's equation, the fundamental equation of fluid mechanics which he formulated around 1755, is nothing but Newton's second law of motion adopted for fluids. This equation states that the acceleration in a portion of fluid is produced by the forces acting there. Now, what can be the typical forces acting inside a fluid? If there are any variations of pressure inside air, we know that air will rush from regions of high pressure to regions of low pressure. So the variation of pressure inside a fluid certainly gives rise to a force. Gravity acting on the fluid will also be another source of force. Euler's equation is a mathematical expression of the fact that acceleration inside a fluid is caused by the joint effects of pressure variations and forces like gravity acting on the fluid.

If you stir water kept in a bucket and leave it, then you will find that after some time the water will come to rest again. That is because there is an internal friction in any fluid like water. When adjacent layers of water move with respect to one another, this internal friction comes into action and tries to brake the motion between adjacent layers, stopping the motions in the bucket of water after a few seconds. This internal friction of fluids is called *viscosity*. Euler's original equation did not include the effects of viscosity. If one applied Euler's equation to study the movements of water in the bucket, one would conclude that the water would never stop and would keep moving forever. It is clear that something is missing from Euler's equation. During the first half of the nineteenth century, scientists like Claude-Louis Navier and George Gabriel Stokes figured out how viscosity can be included in Euler's equation by adding an extra term. When this extra term is included in Euler's equation, the resulting equation is called the *Navier–Stokes equation*. This extra term would make the stirred water in the bucket come to rest a little while after the stirring. Also, if we want to study the flow of water through a horizontal pipe theoretically, it is the Navier–Stokes equation which we have to solve. In fact, many elementary fluid mechanics textbooks discuss how the Navier–Stokes equation can be solved to study the water flow through a horizontal pipe as long as it is laminar.

We now come back to Reynolds's experiment. When water flows sufficiently fast through the pipe, the flow becomes turbulent. But how fast should the water flow be to cause turbulence? Also, we can think of different kinds of liquid flowing through horizontal pipes of different cross-section. If possible, we would like to have some uniform criterion for the onset of turbulence in all such situations. Suppose R is the radius of cross-section of the pipe through which a liquid of viscosity v is moving at speed V. By using the values of R, V and v in a particular situation, we can calculate the value of RV/v, which is called the Reynolds number. By dimensional analysis, it can be shown that this is a pure number without any unit. In other words, whatever system of units one is using to measure R, V and v (the same system of units like cgs or SI has to be used for all three of them), the value of the Reynolds number will turn out to be the same in a particular situation, independent of the system of units. It is found experimentally that, when the Reynolds number for flow through a horizontal pipe exceeds about 3000, the flow becomes turbulent. Thus we do not have to consider the different situations of different liquids flowing through pipes of different sizes. One single result unifies all of these different situations! One can show from the Navier–Stokes equation that the Reynolds number completely determines the nature of the flow. While discussing Alfvén's theorem in Section 4.6, I introduced the dimensionless number called the magnetic Reynolds number. This number happens to be the MHD analogue of the ordinary Reynolds number for fluid mechanics.

By now you must be convinced that turbulence is very pervasive all around us. When we heat water in a pan, we get turbulent motions once convection starts. We have pointed out that water flowing fast through a pipe becomes turbulent. On a windy day, we may find the wind to change its strength and direction continuously, indicating that the air around us is turbulent. Although turbulence is so all-pervasive and some of the greatest physicists of the twentieth century have pondered over it, to this day turbulence remains one of the grand unsolved problems of classical physics. To explain what we mean by turbulence being an unsolved problem, let us again consider the example of water flowing through a horizontal pipe. When a certain amount of pressure is applied, a certain amount of water flows through the pipe per second. We would like to calculate the rate of water flow for a given applied pressure from theoretical principles. As long as the flow is laminar, the rate of water flow can be calculated from the Navier–Stokes equation, the result being known as *Poiseuille's formula*. However, when the flow becomes turbulent, we do not know how to calculate the rate of water flow from first principles. We now have technological capabilities for sending men to the moon, but when water starts flowing quickly through our kitchen tap we do not understand that motion. Apart from a knowledge that the secret of turbulence lies hidden in the Navier–Stokes equation, a complete and satisfactory theory of turbulence has so far eluded physicists. That is why our theoretical understanding of any phenomenon which involves turbulence is rather limited. As we proceed, you will see that turbulence plays a key role in the dynamo process. Dynamo theorists often handle turbulence through rather crude averaging procedures. Until we achieve a breakthrough in our understanding of turbulence, many uncertainties in dynamo theory are bound to remain.

There is one effect of turbulence which is going to be very crucial in our discussion from now onwards. Turbulence is very good at mixing things up. Suppose you put a spoonful of sugar in your coffee and do not stir it. Then it will take ages for the sugar to get mixed in the coffee. But, if you stir the coffee, the sugar gets mixed quickly. The reason is quite simple. You create turbulence in the coffee when you stir it and this turbulence mixes up the sugar. The technical term for this kind of mixing is *diffusion*. Turbulence clearly enhances diffusion. This enhanced diffusion due to the effect of turbulence is called *turbulent diffusion*. We expect that the random gas motions in the sun's convection zone will give rise to turbulent diffusion that would mix various things up in the convection zone.

6.2 Parker's Turbulent Dynamo

Unless you are a particularly unlucky person, you have probably never experienced a tropical cyclone in the open sea. But still you may know that a cyclone involves swirling motions of wind over large distances in

the form of a huge atmospheric vortex. These atmospheric vortices in
the two hemispheres rotate in opposite senses. The reason behind these
vortex-like motions of air is none other than the Coriolis force, which
we discussed at length in the previous chapter and which is behind Joy's
law of the tilts of sunspot pairs. It is the Coriolis force which deflects air
moving over large regions into quasi-circular paths.

Now we come to Parker's famous idea of how the toroidal magnetic field may give rise to the poloidal magnetic field. Consider a
hot blob of gas rising in the sun's convection zone. The Coriolis force
due to the sun's rotation will make this rising blob rotate as it rises,
exactly like cyclones in the earth's atmosphere. As a result, the blob
will have a corkscrew-like motion—rotating while rising. The mathematical term for this kind of motion is 'helical'. Suppose a part of the
toroidal magnetic field inside the sun has got caught in a helical rising hot blob of gas. Because of Alfvén's theorem of flux freezing, this
magnetic field will be dragged by the rising hot blob and will also get
twisted by the helical motion. This is indicated in Figures 6.3(a) and
6.3(b). The first figure simply shows a toroidal magnetic field which has
been produced from a poloidal magnetic field by differential rotation.
In the second figure, you can see that a part of the magnetic field lies in
the poloidal plane (i.e. a plane containing the rotation axis of the sun)
in the form of a magnetic loop after being twisted by the helical motion
of the hot blob. While a magnetic loop will not always lie completely
in the poloidal plane, there will generally be a projection of the loop
in the poloidal plane. Just as cyclones in the two hemispheres of the
earth rotate in the opposite sense, one can show that the helical motions of hot gas in the sun's convection zone will have opposite senses
in the two hemispheres. If you keep in mind that the toroidal magnetic
field in the two hemispheres also has opposite signs, then you can easily arrive at the conclusion that the projected loops of magnetic field in
the poloidal plane will have the same sense in the two hemispheres, as
indicated in Figure 6.3(b).

Figure 6.3(c) shows many such magnetic loops in the poloidal
plane—all having the same sense. We certainly have strong turbulent
diffusion in the sun's convection zone, which will mix things up. We
expect the turbulent diffusion to blend the several magnetic loops
together and produce the large-scale magnetic field shown by the
dashed line in Figure 6.3(c). The upshot of the whole process is that,
starting from the toroidal magnetic field shown in Figure 6.3(a), we

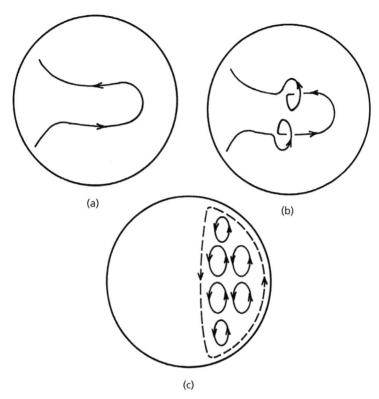

Figure 6.3 Different stages of the dynamo process. See text for explanations.

have ended up with the poloidal magnetic field indicated by the dashed line in Figure 6.3(c).

Figure 6.4 schematically summarizes the main idea behind Parker's turbulent dynamo. The poloidal and the toroidal magnetic fields sustain each other through a feedback loop. Starting from the poloidal field, we can get the toroidal field due to stretching by differential rotation. The toroidal field, in its turn, can give rise to the poloidal field by helical turbulence, as we have discussed above. Take another look at Figure 2.11 showing the oscillation between the toroidal and poloidal magnetic fields. As I mentioned in our discussion of this figure, this oscillation could be established from observational data only many years after Parker's work on the turbulent dynamo. However, Parker could envisage such an oscillation and the scheme encapsulated in Figure 6.4

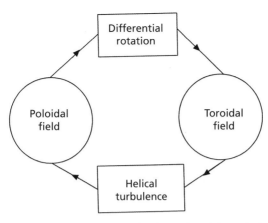

Figure 6.4 Schematic representation of Parker's idea of the turbulent dynamo.

shows how such an oscillation might arise. As a result of this oscillation, we expect to have a periodic cycle, which presumably offers the theoretical explanation of the sunspot cycle. This is, in a nutshell, the main idea behind Parker's theory of the turbulent dynamo. Do not worry if you are not immediately comfortable with Parker's idea. It is a rather unconventional idea and most of the scientists working in this field certainly did not feel comfortable with this idea when it was first proposed.

It is unanimously acknowledged that Parker's 1955 paper is the most important landmark in the historical development of dynamo theory.[2] However, no matter how wonderful an idea is, physicists do not take it seriously until detailed mathematical calculations show that it really works. Parker also did not just stop with the idea. A major part of his 1955 paper is devoted to translating his ideas into mathematical equations and showing that these mathematical equations really have solutions resembling the sunspot cycle. In a turbulent situation, velocity and magnetic fields are expected to change rapidly with time and space. So the best bet is to derive an equation for the average magnetic field. Parker managed to obtain an equation for the average magnetic field, which is now called the *dynamo equation*.

After deriving the equation for the average magnetic field, Parker was able to obtain some solutions of this equation. We have discussed Cowling's anti-dynamo theorem that a symmetric solution of the dynamo problem is not possible. Now, within the sun's convection zone, the turbulent velocities are not symmetric and they distort magnetic fields also in ways which are not symmetric. So we get around Cowling's theorem. This theorem does not exclude symmetric solutions of the average magnetic field from the dynamo equation, and symmetric solutions are found to exist. Some of the earlier scientists like Walter Elsasser and Edward Bullard who attempted to demonstrate the dynamo process with laminar flows (without success) had to do very complicated mathematical calculations because they knew that simple symmetric solutions were excluded by Cowling's theorem. Interestingly, the dynamo equation for the average magnetic field turned out to be much easier to handle because symmetric solutions now became possible.

Parker showed that the dynamo equation allows a wave-like solution. What is the significance of that? Let us consider a simple wave on the surface of water. We see that a crest of the wave keeps moving. Now, in a wave obtained from the dynamo equation, the crest corresponds to a region where the toroidal magnetic field is strong. We expect many sunspots to form from a strong toroidal magnetic field, which means that the crest of the dynamo wave would be a region where many sunspots appear. In Section 2.3, I have discussed that sunspots appear at lower and lower latitudes with the progress of the sunspot cycle. A natural explanation for this is that there is a dynamo wave underneath the sun's surface propagating equatorward, so that the crest moves to lower and lower latitudes with time. Parker solved the dynamo equation to discover that such a dynamo wave follows from mathematical theory, providing the first possible explanation of the sunspot cycle. In the case of the water wave, a crest is usually followed by a trough where the water level becomes lower than normal. Exactly in the same way, if the toroidal magnetic field in one cycle of the dynamo wave is in a particular direction, it is in the opposite direction in the next cycle. This explains why the magnetic polarities of sunspot pairs (as seen in a magnetogram map like Figure 2.7) reverse from one cycle to the next.

One important question connected with the dynamo wave is: what determines its direction of propagation? For example, why don't we have the dynamo wave propagating towards the pole, making sunspots appear at higher and higher latitudes with the progress of the sunspot cycle? Parker found that the differential rotation and the helical turbulence (which together produce the dynamo wave) have to satisfy a certain mathematical relation to ensure that the dynamo wave propagates equatorward. A few years later, Hirokazu Yoshimura established this mathematical relation for a more general situation.[3] We call it the *Parker–Yoshimura sign rule*. Since this is a slightly technical topic, I will not go into any more detail. Let me just mention that the dynamo wave propagates in the correct direction if the nature of the differential rotation and the helical turbulence is such that the Parker–Yoshimura sign rule is satisfied (this rule is written down in Appendix H).

While today we recognize Parker's 1955 paper on the dynamo as an extraordinary scientific achievement, its importance was realized rather slowly. One easy way of gauging whether a paper is considered to be important by the scientific community is to note how often other scientists cite the paper in their works. In the first 10 years after its publication, Parker's paper barely picked up about 15 citations! One reason for this neglect was probably the fact that Parker's derivation of the dynamo equation was rather heuristic and not very satisfactory. In 1957 Cowling wrote the famous textbook *Magnetohydrodynamics*, from which a whole generation of students learnt the subject. Cowling made the following comment about Parker's dynamo work in his book:

> The argument is not altogether satisfactory; a more detailed analysis is really needed. Parker does not attempt such an analysis; his mathematical discussion is limited to elucidating the consequences if his picture of what occurs is accepted. But clearly his suggestion deserves a good deal of attention.[4]

The kind of analysis Cowling asked for was finally provided in 1966 by a group of East German scientists, as we shall discuss in the next section. Their work at last put the subject of turbulent dynamo on a firmer foundation. Interestingly, the second edition of Cowling's book appeared in 1975 when the importance of Parker's 1955 paper was generally recognized and Cowling removed the comment quoted above from the second edition.

At the time of Parker's work on the dynamo, the study of turbulence was in its infancy. We usually think of turbulence as something which mixes things up and creates disorder in the fluid. An amazing aspect of Parker's theory of the turbulent dynamo is the suggestion that an ordered magnetic field can actually arise out of turbulence. This is called *inverse cascade* in technical jargon. To the best of my knowledge, Parker's work provided the first example of inverse cascade in turbulence. With further research on turbulence, scientists have gradually become aware of many other situations where inverse cascade takes place and ordered structures arise out of turbulence. The study of inverse cascade has become a very big and important topic in turbulence research.

While discussing our work on magnetic buoyancy in the previous chapter, I mentioned that all dynamo theorists in the 1980s held the view that the toroidal magnetic field inside the sun could not be stronger than 1 tesla. Now I can give the reason behind this view. One important step in Parker's dynamo is the twisting of the toroidal magnetic field by the helical turbulence, as seen in Figure 6.3(b). If the toroidal field is very strong, then its magnetic tension would be so large that the helical turbulence would be unable to twist it. One can estimate the strength of helical turbulence in the sun's convection zone, from which it follows that the toroidal magnetic field cannot be stronger than 1 tesla if it has to be twisted by helical turbulence. As I mentioned towards the end of Chapter 5, the study of buoyant rise of flux tubes by us and by other groups established that the toroidal magnetic field has to be as strong as 10 tesla. Then helical turbulence would be unable to twist such a strong toroidal magnetic field. Parker's model of the sunspot cycle, in spite of its great historical importance, is presumably not the correct model. I shall describe in the next chapter what many of us now consider to be the correct model of the sunspot cycle.

6.3 The Oeuvre from East Germany

The group in East Germany which provided a systematic derivation of the dynamo equation was headed by Max Steenbeck. An account of his life reads almost like something taken out of the pages of a story book. Here I merely give a brief outline of his life. His original training was in experimental work on X-rays, electrical discharge through gases and nuclear physics. One of his early achievements was to design and build a device for accelerating electrons known as the betatron. He was still working for a private company in Berlin when World War II broke out. Steenbeck was captured by the Soviet Red Army when they occupied a part of Berlin at the end of World War II and was sent to a detention camp in Poland, where Steenbeck was close to starvation. However, Stalin wanted the Soviet Union to make the atom bomb and knew that any captured German physicist with a knowhow of experimental nuclear physics would be a very valuable asset in this project. So Steenbeck was brought to a secret place near the Black Sea, where he was to work on the Soviet atom bomb project. Although his personal

Figure 6.5 Max Steenbeck (1904–1981) who, along with his younger colleagues Fritz Krause and Karl-Heinz Rädler, provided a formal mathematical formulation of dynamo theory.[5]

freedom was restricted and he was not allowed to travel, his family could join him and he at least had a comfortable life, working there for several years. Shortly after Stalin's death, Khrushchev initiated a period of political openness and Steenbeck was at last allowed to go home in 1956, after spending 11 long years in the Soviet Union. Although Steenbeck had the option of settling in West Germany, he decided to settle in the communist East Germany. Even after spending several years in Stalin's Soviet Union, Steenbeck still had the firm belief that communism held the best promise for eliminating all the evils of human society. For several years, Steenbeck was the Director of the Institute for Magnetohydrodynamics in Jena. At that time, it was still thought that the commercial production of energy through nuclear fusion would be possible in the near future. That was the main goal of Steenbeck's institute.

While directing research on the nuclear fusion project, Steenbeck also became interested in the theoretical question of how magnetic fields arise in astronomical bodies. Being an experimental scientist himself, Steenbeck wanted to collaborate with some younger colleagues who were trained in theoretical research. Fritz Krause, trained as a mathematician, and Karl-Heinz Rädler, trained as a physicist, joined Steenbeck in his investigation of the dynamo problem. The landmark paper by Steenbeck, Krause and Rädler putting the foundations of dynamo theory on a firm footing appeared in 1966.[6] Although English was already becoming the international language of science, Max Steenbeck—proud German that he was—could not think of publishing his research in English. The landmark paper of 1966 as well as the subsequent papers from the group were all written in German and appeared in German journals. The language barrier initially came in the way of these papers getting the full attention they deserved. But gradually scientists working in this field around the world started realizing that something tremendously important was happening in East Germany. Two distinguished dynamo theorists—Paul Roberts and Michael Stix—translated several papers from this group into English in 1971 and put them together in a volume. To the best of my knowledge, these translations are still not available over the internet. Somebody ought to take the initiative of putting these translations on the internet and linking them to other important internet sites through which they can be tracked easily. Many years ago, I somehow managed to get hold of a copy of the volume containing these translations. Over the years, I have zealously guarded this volume, refusing colleagues and friends who have wanted to borrow it for consultation. I know that this volume could not be replaced if it were lost.

What did Steenbeck and his younger colleagues achieve? Why is their work regarded as a landmark in the development of dynamo theory? Since their work is of a rather technical nature, it is not easy to answer these questions in a popular science book. I have already said that Parker's original derivation of the dynamo equation was rather heuristic. Steenbeck, Krause and Rädler provided a more systematic derivation, putting the dynamo equation on a firmer foundation. While deriving the dynamo equation, they came up with a systematic mathematical formulation for the whole subject known as *mean field magnetohydrodynamics*. Any serious student of dynamo theory has to master this formulation. Although it is difficult to explain this work in a book without mathematics, let me make a few remarks.

I have pointed out that one of the central results of MHD is Alfvén's theorem of flux freezing, which follows the principle of electromagnetic induction applied to plasmas (Section 4.6). When one combines the basic equation of electromagnetic induction with Ohm's law for plasmas, one gets an equation known as the *induction equation*, the central equation of MHD. A discussion of this equation is presented in

Appendix D. This equation tells us how the magnetic field in a plasma changes with time. When there are motions inside the plasma, the magnetic field lines tend to get dragged, as indicated in Figure 4.10(c). At the same time, the resistance of the plasma tries to cause a decay of the currents through the plasma and hence the magnetic field. The induction equation incorporates both these effects of the magnetic field: the tendency to be dragged by plasma motions and the decay due to resistance. When the decay due to resistance is not so important (which is technically the situation in which the magnetic Reynolds number is high), the induction equation leads to Alfvén's theorem of flux freezing. What Steenbeck, Krause and Rädler did was to carry out a mathematical analysis of the induction equation in a situation of turbulence.

Let us come back to Reynolds's experiment on the flow of water through a pipe discussed at the beginning of this chapter. Even when water flows sufficiently fast and turbulence sets in, we would always find an average velocity of water corresponding to a flow downstream. However, if we measure the velocity at a particular point inside the water, we would find that the velocity will not just be this average velocity. But, added to this average velocity, there would be another part of velocity fluctuating randomly with time. It is this randomly fluctuating part of the velocity which causes the dye to spread out in a turbulent situation, as seen in Figure 6.2. In the presence of turbulence that exists in the sun's convection zone, Steenbeck and his younger colleagues realized that both the magnetic field and the velocity at a point can be split into two parts: the average part and a part fluctuating randomly. Then they did some calculations with the induction equation assuming certain statistical properties of the fluctuating parts of the magnetic field and the velocity. Through a systematic procedure, they arrived at the dynamo equation, the central equation in dynamo theory, of which a version had earlier been derived by Parker through heuristic arguments. As the quotation from Cowling in the previous section shows, many scientists did not find Parker's arguments fully satisfactory and that might have been one of the reasons why Parker's 1955 paper on the dynamo process did not initially receive enough attention. When the Steenbeck, Krause and Rädler 1966 paper showed how the dynamo equation can be derived systematically by considering the induction equation in a turbulent situation, the whole subject was put on a more logical foundation. Readers with a knowledge of vector analysis will find a brief outline of the subject in Appendix H.

Figure 6.3 should make it clear that helical motions of the turbulent plasma are crucial in twisting the toroidal magnetic field to generate the poloidal magnetic field. The dynamo equation was indeed found to have a term with a coefficient corresponding to these helical motions. Steenbeck, Krause and Rädler used the Greek letter α to denote this coefficient. Hence the process of the poloidal magnetic field generation from the toroidal magnetic field by helical motions is called the α-effect. It lies at the heart of classical dynamo theory. The lower part of the loop shown in Figure 6.4 corresponds to this α-effect. Since the usual symbol for rotation in mathematical theory is the Greek letter Ω, the upper part of the loop corresponding to differential rotation generating the toroidal magnetic field from the poloidal magnetic field is called the Ω-effect. A theoretical dynamo model based on the ideas encapsulated in Figure 6.4 is called an $\alpha\Omega$ dynamo. While the pioneers of dynamo theory assumed that the sun's magnetic cycle is caused by an $\alpha\Omega$ dynamo, it was clear that the earth's magnetic field has to be generated somewhat differently, because there is not much differential rotation inside the earth. The earth rotates almost like a solid body. The helical motions of the plasma (the α-effect) can also twist the poloidal magnetic field to produce the toroidal magnetic field. We believe that the toroidal field generation from the poloidal field inside the earth proceeds by the α-effect, unlike the sun within which differential rotation does this job. The kind of dynamo operating inside the earth is called an α^2 dynamo in the technical jargon of this subject. In a schematic representation of the earth's dynamo like Figure 6.4, one would have to put 'Helical turbulence' in both of the boxes at the top and the bottom. For readers desirous of knowing more about how the earth's magnetic field is produced by the dynamo process, I shall present some discussion in Chapter 10.

While discussing Parker's work on the dynamo, I mentioned that he had obtained a solution of the dynamo equation corresponding to a propagating wave, which was proposed as a theoretical model of the sunspot cycle. Parker used rectangular coordinates in infinite space to do the first simple calculations. In 1969 Steenbeck and Krause solved the dynamo equation within a spherical region, taking the spherical surface of the sun into account.[7] These calculations were more complicated and had to be done with a computer. Steenbeck and Krause assumed that sunspots would form in a region of the sun whenever the toroidal magnetic field there is sufficiently strong. We have seen in Figure 2.3

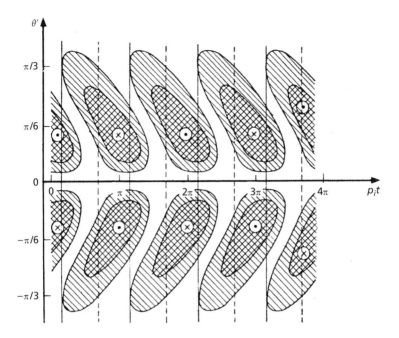

Figure 6.6 The first theoretical butterfly diagram obtained by Steenbeck and Krause in 1969.

that observations of sunspots at different latitudes at different times lead to what is called a butterfly diagram. From their theoretical model, Steenbeck and Krause were able to construct a theoretical butterfly diagram, which is shown in Figure 6.6. Constructing such a theoretical butterfly diagram was indeed a landmark in the development of solar dynamo theory.

6.4 The End of the Road?

Although I have been working on dynamo theory for many years, I am embarrassed to say that I do not know who first coined the name 'dynamo theory'. Most working scientists are notorious for not being very history-conscious. They usually have only a rather hazy idea of what happened in their research fields before they started their research careers. It should be possible to dig up the history of the name 'dynamo theory' by perusing the old literature and somebody should do it some day! I can only say that the dynamo process for generating magnetic fields in astronomical bodies has certain similarities with an ordinary electromagnetic dynamo, which is the source of electrical energy we receive in our homes. In an ordinary dynamo, we have a conducting coil which is made to rotate in a magnetic field, cutting magnetic flux lines and thereby producing currents by Faraday's law of electromagnetic induction. We do not have discrete coils inside an astronomical body, but we do have rotating blobs of plasma which are also good conductors of electricity. It is again the phenomenon of electromagnetic induction which is responsible for driving currents through these rotating plasma blobs and giving rise to magnetic fields.

The theoretical butterfly diagram obtained by Steenbeck and Krause in 1969 motivated several scientists to build their own theoretical models of the solar dynamo in the 1970s. Michael Stix, Paul Roberts and Hirokazu Yoshimura were amongst the scientists who made important contributions in this field. One crucial input in a theoretical solar dynamo model is the differential rotation of the sun. As I discussed in Section 3.7, helioseismology started providing information about differential rotation from the mid-1980s. Nothing much was known about differential rotation in the interior of the sun in the 1970s when the first detailed models of the solar dynamo were being constructed. Scientists working in this field would assume some kind of differential rotation which seemed reasonable to them. One requirement for solar dynamo models is that the dynamo waves should propagate towards the equator. I have already mentioned that a condition called the Parker–Yoshimura sign rule has to be satisfied for this to happen. All the scientists would choose the differential rotation in such a way that the Parker–Yoshimura sign rule was satisfied. Since theoretical butterfly diagrams such as the one shown in Figure 6.6 resembled the observational butterfly diagram so well, everybody was confident that solar dynamo theory was on the right track.

A first jolt to this confidence came when helioseismology succeeded in mapping the differential rotation in the interior of the sun. We have shown this in Figure 3.9. The real differential rotation turned out be completely different from what dynamo theorists in the 1970s were assuming. Then the big blow came from the magnetic buoyancy calculations which my co-workers and I initiated, as described in Chapter 5. At that time, most of the solar physicists had so much confidence in the existing dynamo models that it was inconceivable that the toroidal magnetic field could be stronger than 1 tesla, since helical motions of the plasma would be unable to twist stronger toroidal magnetic fields to produce the poloidal magnetic field. Initially we had difficulty in believing our own results when we started finding that the toroidal magnetic field strength has to be closer to 10 tesla. When independent calculations by several groups confirmed that the toroidal magnetic field must be of the order of 10 tesla, it became clear that totally new kinds of solar dynamo models were needed. From the 1990s serious efforts began to construct models of the solar dynamo using the differential rotation measured by helioseismology which would be able to account for the much stronger toroidal magnetic field. I am lucky that I could be personally involved in these efforts. We actually had to go back to some early ideas about the solar dynamo which were developed by Horace Babcock and Robert Leighton in the 1960s, but the significance of which was not fully appreciated at that time. We shall describe these developments in the next chapter.

Before I end this chapter, I would like to mention that Max Steenbeck, having begun his career as an experimental physicist, was always very interested in laboratory demonstrations of the dynamo process. Certainly dynamos are used by electricity companies for the generation of electricity. But, in a terrestrial laboratory, can we have a self-excited fluid dynamo like the dynamos inside astrophysical bodies? Any electrically conducting fluid obeys MHD equations. Can we make an electrically conducting fluid (like molten metal) flow in such a way that a magnetic field is sustained inside it as long as the flow continues? Literally, can we bring down the dynamo from the heavens to the earth? It should be clear from our discussion that Alfvén's theorem of flux freezing lies at the heart of the dynamo process. I pointed out in Section 4.6 that this theorem is valid for systems with high magnetic Reynolds number (i.e. essentially systems of very large size). In laboratory experiments, it is difficult to achieve sufficiently high magnetic Reynolds numbers and hence Alfvén's theorem of flux freezing holds only approximately. It is certainly not easy to have a self-excited dynamo due to fluid motions in a laboratory setup. Although Steenbeck initiated such laboratory experiments, the first success came several years after his death. Due to the influence of Steenbeck, a tradition of doing such experiments developed in

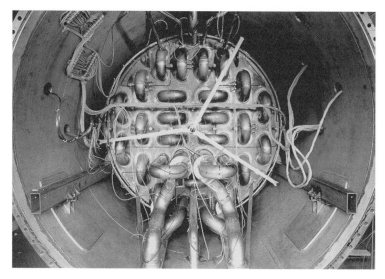

Figure 6.7 The interior of the vessel used in the Karlsruhe dynamo experiment. The diameter of the vessel is about a metre. Credit: K.-H. Rädler.

Eastern European countries. In 2000 Agris Gailitis's group, working with molten sodium in Riga, Latvia, succeeded in achieving a self-excited fluid dynamo in the laboratory for the first time.[8] Since that time, a few more such experiments have been set up in different countries. Figure 6.7 shows the interior of the vessel used in the Karlsruhe dynamo experiment.

7

The Conveyor Belt inside the Sun

7.1 The Death of Sunspots and the Aftermath

A large sunspot typically lives for about 10 days before disintegrating and disappearing. Why do sunspots die? One can give a philosophical answer that all things in the world are impermanent. But that is the kind of answer which does not usually satisfy physicists. So let us address the question in some more detail: why do sunspots die and what determines their lifetimes?

We have discussed turbulent diffusion in Section 6.1. When a lump of sugar is put in your coffee and you create turbulence in the coffee by stirring it, the sugar disperses throughout the coffee. Now, a sunspot is a lump of magnetic field floating on the turbulent convection zone of the sun. We expect turbulent diffusion to disperse the magnetic field of the sunspot. Careful observational studies of the decay of sunspots support that this is what happens. The magnetic field initially bundled up inside the sunspot becomes dispersed over a few days due to the effect of turbulent diffusion in the sun's convection zone. Eventually the concentration of magnetic field ceases to exist and the sunspot disappears.

That sunspots often appear in pairs having opposite polarities has been repeated several times. One important law pertaining to sunspot pairs is Joy's law about their tilts, which was discovered by Alfred Joy in 1919 and which was explained theoretically by D'Silva and me nearly three-quarters of a century later. According to this law, one sunspot will be nearer the equator and the other sunspot further away. Figure 7.1(a) shows a tilted sunspot pair. For the sake of illustration, we have taken the sunspot nearer the equator to be positive. Let us now consider what happens when such a tilted sunspot pair decays. We expect the positive magnetic field to get dispersed over the region indicated by '+' signs in Figure 7.1(a). Similarly the negative magnetic field will be dispersed over the region indicated by '—' signs. The result is a positive polarity patch at a lower latitude and a negative polarity patch at a higher latitude. Figure 7.1(b) shows what the magnetic field lines should look like,

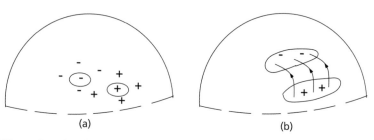

Figure 7.1 Sketches depicting the Babcock–Leighton mechanism for the production of the poloidal magnetic field from the decay of a sunspot pair. See text for explanation.

connecting these two patches of opposite polarity. These magnetic field lines would have a significant projection in the poloidal plane. We thus come to the conclusion that a tilted sunspot pair is an intermediate step in the process of converting the toroidal magnetic field to the poloidal magnetic field. As we have discussed in detail in Chapter 5, the tilted sunspot pair arises from the toroidal magnetic field of which a part rises due to magnetic buoyancy. Once the tilted sunspot pair decays, we get some poloidal magnetic field from it.

The idea outlined above was developed in two influential scientific papers—the first one by Horace Babcock in 1961[1] and the second one by Robert Leighton in 1964.[2] Hence this mechanism of getting a poloidal magnetic field starting from a toroidal magnetic field is called the *Babcock–Leighton mechanism*. In the present-day research world of physics, there is essentially a non-overlapping division between two kinds of scientists: those who do theoretical research and those who do experimental research. In our age of specialization, very few brave individuals succeed in straddling both sides of the divide. One often expects major theoretical advances in a theoretical subject to be made by theoretical scientists. It is a curious fact that things did not happen quite this way in the early years of dynamo research. In Section 6.3, I discussed the very important contributions of Max Steenbeck, who was not a theoretical scientist. Now we are discussing the influential ideas developed by Babcock and Leighton. Neither of them was a primarily theoretical scientist. Both Babcock and Leighton were very good at building instruments. I have often wondered why some of the seminal contributions in the early years of dynamo research came from men who were not primarily theoretical scientists. Some of

Figure 7.2 Horace Babcock (1912–2003), who proposed an influential idea to explain how the sunspot cycle arises.

these seminal works required more intuition than formal theoretical reasoning. Perhaps people who are good at building instruments have better intuition than those who had been formally trained for theoretical research. With increasing specialization, it is becoming rarer to find an instrument-builder who has a good feeling for theoretical ideas. But this breed of scientists has not become totally extinct yet. I have a good Japanese friend: Saku Tsuneta. He played a key role in some of the most successful recent space missions to study the sun. I had several opportunities of discussing science with him when I was a visiting professor in Tokyo. Whenever I discussed theoretical topics with him, I was amazed by his grasp of theoretical ideas.

Robert Leighton and his students were the first to discover that a point on the sun's surface continuously goes up and down, ultimately leading to the founding of the new discipline helioseismology. While introducing this topic in Section 3.6, I have said a few words about Leighton, mentioning that he was the second author of the celebrated *Feynman Lectures*. Let me now say a few words about Horace Babcock. He was the son of the distinguished astronomer Harold Babcock, one of the pioneers in developing the magnetograph to study magnetic fields at the solar surface outside sunspots. Horace began his research career by collaborating with his father Harold. I have already mentioned in Section 2.5 that in 1955—the year in which Parker published his famous paper on dynamo theory—the father-and-son team provided the first decisive evidence of the magnetic field in the sun's polar region. Later Horace Babcock charted out a course of his own. One of his most important contributions to astronomy was to propose the idea of what is called *adaptive optics*, a technique for improving the image quality of telescopes by correcting for atmospheric disturbances. Horace Babcock was several decades ahead of his time when he proposed this technique. Only after the development of fast computers, could this technique be realized. Now it is used routinely in all large telescopes around the world.

The works of Babcock and Leighton on the generation of the sun's poloidal field were rather heuristic and did not attract much attention for several years. After Steenbeck, Krause and Rädler systematically developed the theory of the α-effect earlier proposed by Parker, most of the detailed dynamo calculations were based on the α-effect. As I mentioned in the previous chapter, this effect requires the twisting of the toroidal magnetic field by helical turbulence and would not work in the sun's convection zone if the toroidal magnetic field is much stronger than 1 tesla. The simulations of magnetic buoyancy described in Chapter 5 suggested that the toroidal magnetic field at the bottom of the sun's convection zone is more likely to be around 10 tesla. Since the α-effect would be completely suppressed for such a strong toroidal magnetic field, some of us started wondering whether the old ideas of Babcock and Leighton could be invoked to explain how the poloidal magnetic field of the sun is produced. This would require a modification of the scheme shown in Figure 6.4 suggesting how the solar dynamo works. The toroidal magnetic field would still be produced from the poloidal magnetic field by the action of differential rotation. But, in order to get back the poloidal magnetic field from the toroidal magnetic field, we would now use the Babcock–Leighton mechanism instead of helical turbulence. Could the solar dynamo work this way?

I have already discussed in Section 6.2 that one of the requirements of solar dynamo theory is to obtain an equatorward propagating wave of the toroidal magnetic field to explain why sunspots appear at lower and lower latitudes with the progress of the sunspot cycle. I mentioned that a certain condition known as the Parker–Yoshimura sign rule has to be satisfied for this. When the first detailed models of the solar dynamo were being constructed in the 1970s (based on the α-effect), nothing was known about the distribution of angular velocity in the interior of the sun's convection zone. So different persons working on solar dynamo models could assume different distributions of angular velocity in the solar interior, subject only to the condition that the Parker–Yoshimura sign rule should be satisfied. When helioseismology eventually mapped the distribution of the sun's angular velocity, it was found that this distribution, shown in Figure 3.9, was totally different from what most of the dynamo modellers had assumed. By the mid-1990s when we realized that the α-effect may not work for the sun, the distribution of the sun's angular velocity was more or less completely pinned down by helioseismology. Some of us who started toying with the idea that the sun's poloidal magnetic field might be generated by the Babcock–Leighton mechanism (if the α-effect does not work) had the following question before us: will the Babcock–Leighton mechanism combined with the differential rotation measured by helioseismology satisfy the Parker–Yoshimura sign rule? A simple check showed that the answer to this all-important question was negative. If the poloidal magnetic field of the sun was produced as per the Babcock–Leighton mechanism, it followed that the dynamo wave would propagate poleward and would cause sunspots to appear at higher and higher latitudes with the progress of the sunspot cycle, completely contradicting observations. At first sight, it appeared that the Babcock–Leighton mechanism was a doomed idea and would not work. Since it was clear by that time that the α-effect would not work for the sun, some of us were desperately trying to build solar dynamo models using the Babcock–Leighton mechanism. Could there be some way of salvaging this mechanism? Could we possibly have overlooked something? I got involved in answering these questions along with some colleagues. We turn to this topic now.

7.2 The Missing Link: A Mysterious Flow

Suppose there are several sunspot pairs like the one shown in Figure 7.1(a) on the solar surface—all of them having the negative polarity sunspot at the higher latitude. When they all decay, we expect

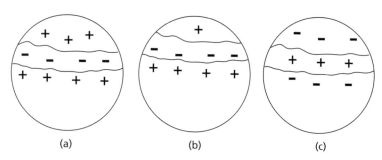

Figure 7.3 Sketches indicating how the magnetic field on the sun's surface outside sunspots evolves with time.

to have a band of negative magnetic polarity at the high latitude and a band of positive magnetic polarity at the low latitude. This is shown in Figure 7.3(a). As magnetograms improved and solar astronomers were able to make accurate measurements of magnetic fields on the solar surface outside sunspots in the 1960s and 1970s, it was indeed found that there are such bands on the solar surface having a certain magnetic polarity.

It was discovered by Carrington and Spörer in the nineteenth century that sunspots appear at lower and lower latitudes with the progress of the sunspot cycle. In the case of these bands of magnetic fields outside sunspots, the surprising discovery was that the bands kept shifting to higher and higher latitudes, in contrast to the sunspots. Figure 7.3(b) depicts how the bands shown in Figure 7.3(a) would look after a few years. Then Figure 7.3(c) shows their appearance still a few years later. When the band of negative magnetic polarity reaches the pole which earlier had positive magnetic polarity, the sun's polar magnetic field gets reversed. You can see such polar reversals in Figure 2.11 which showed how polar magnetic fields of the sun varied with time.

Why do these bands of magnetic field outside sunspots drift to higher latitudes, while sunspots appear at lower and lower latitudes with the progress of the sunspot cycle? A clue is provided by Alfvén's theorem of flux freezing, which tells us that a magnetic field can be carried with a plasma flow. Suppose the plasma near the sun's surface keeps moving continuously from the equator to the pole. Then that plasma would carry the magnetic field with it and this would explain why the bands of magnetic polarity drift to higher latitudes with time. From various kinds of measurements, it is now established beyond doubt that there is

such a plasma flow near the sun's surface, carrying the bands of magnetic polarity along with it. This flow is called the *meridional circulation* of
the sun. The maximum value of this flow at mid-latitudes is about 70
kilometres per hour—the typical speed with which a car moves along
a highway.

We certainly do not expect the plasma to pile up near the sun's poles.
So the plasma which has flown to the pole must be brought back again
to the equator by some path underneath the sun's surface. We still do
not have a very good understanding of how the meridional circulation
is produced. The turbulence in the sun's convection zone gives rise to
some stresses and it is believed that these stresses cause the meridional
circulation. If that is the case, then the meridional circulation would be
confined completely within the sun's convection zone and we would
expect the return flow from the pole to the equator to take place somewhere deeper down within the convection zone. The most plausible ex-
pectation is that the return flow is at the bottom of the convection zone.
Figure 7.4 shows a sketch of the expected flow pattern of this meridional
circulation. Basically we have something like a gigantic conveyor belt in
the sun's convection zone. This conveyor belt carries everything from
the equator to the poles in the upper layers of the convection zone.

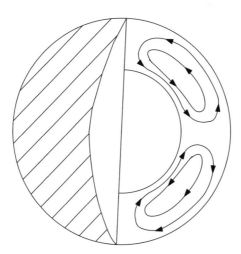

Figure 7.4 The expected flow lines of the meridional circulation inside the
sun's convection zone: the conveyor belt.

Then the lower part of the conveyor belt carries things back from the poles to the equator in the lower layers of the convection zone.

Does such a conveyor belt—the meridional circulation, in more scientific language—really exist inside the sun? Over the years, we have got more and more evidence of its existence. I have discussed how helioseismology has been used to map the angular velocity inside the sun, leading to the remarkable Figure 3.9. Helioseismology has also been used to map the meridional circulation underneath the sun's surface. However, the results become more and more uncertain as we go deeper. Although there has to be a return flow from the pole to the equator to make sure that the matter brought to poles through surface layers does not keep accumulating there, so far there are no unambiguous measurements of this return flow. So we have to depend on theoretical arguments to infer the nature of this return flow. We know that the amount of matter carried equatorward by the return flow has to be equal to the amount of matter carried poleward through the upper layers. Otherwise, matter will keep accumulating somewhere. Now, the gas at the bottom of the sun's convection zone is much denser than the gas near the surface. So the speed of the return flow has got to be much less than the speed of the poleward flow near the surface, if the return flow in the deeper layers has to carry exactly the same amount of matter equatorward as what is carried poleward through the upper layers. Such arguments suggest that the speed of the return flow at the bottom of the sun's convection zone is likely to be not more than 10 kilometres per hour.

This is a remarkable result for the following reason. I have pointed out in Section 6.2 that the latitudinal drift of sunspots can be explained on the basis of a dynamo wave propagating equatorward. From observations of sunspots, one can make an estimate of the speed of this dynamo wave. This also turns out to be somewhat less than 10 kilometres per hour! Is it mere coincidence that the estimated speed of the meridional circulation at the bottom of the sun's convection zone turns out to be essentially the same as the required speed of the dynamo wave? Most physicists do not like such coincidences. Many of us hold the view that this is not mere coincidence. The strong toroidal magnetic field of the sun is produced at the bottom of the convection zone by the strong differential rotation present there and is then carried equatorward by the meridional circulation. If sunspots are produced in the region where the toroidal magnetic field is particularly strong, then this region will keep shifting equatorward at the same speed at which the meridional circulation is moving with the toroidal magnetic field. Now, I have already mentioned that a solar dynamo model in which poloidal magnetic field generation is due to the Babcock–Leighton mechanism is expected to give rise to a poleward propagating dynamo wave according to the Parker–Yoshimura sign rule. This sign rule was derived without including the meridional circulation in the theoretical analysis. If there is a meridional circulation, is it possible to get around this sign rule by making the meridional circulation carry the dynamo wave with it? This was the crucial question which we had to address when some of us started wondering in the early 1990s whether the Babcock–Leighton mechanism could be the mechanism for generating the poloidal magnetic field if the α-effect could not be operational on a very strong toroidal magnetic field. But let me say a few things more about the magnetic fields outside sunspots before we come to this crucial question.

By now you must feel completely at home with the butterfly diagram of sunspots. The first such diagram constructed by Maunder was shown in Figure 2.3. This is essentially a diagram involving latitude plotted against time. It is possible to represent the magnetic field outside sunspots also in this diagram. Suppose we consider a latitude of the sun at a certain time. If we have the values of magnetic field in all longitudes (in the visible hemisphere of the sun) at this latitude, then we can obtain a longitude-averaged value of the magnetic field at this latitude. The longitude-averaged values of the magnetic field outside sunspots at different latitudes at different times can be represented in the butterfly diagram along with sunspots by using a grey scale. Figure 7.5 shows such a diagram. The grey scale in the background indicates the values of the magnetic field outside sunspots (see the bar next to the plot for the values) at different latitudes at different times. You can clearly see from this plot that the magnetic field outside sunspots drifts poleward with the solar cycle, in contrast to the sunspots which appear closer and closer to the equator with the progress of the cycle.

Let me now make some remarks about the value of the magnetic field outside sunspots. The bar in Figure 7.5 gives the values of the magnetic field corresponding to the grey scale. The unit is gauss, which is 10^{-4} of a tesla. You can see that this magnetic field is strongest near the poles where it has values close to 10 gauss, which is 0.001 tesla. Remember that the magnetic fields inside sunspots have a much higher value of

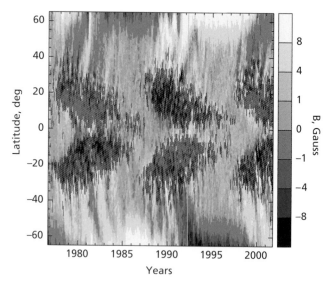

Figure 7.5 A butterfly diagram in which, apart from the sunspots, the magnetic field outside sunspots evolving in latitude and time is shown by the grey scale.

about 0.3 tesla. Now, there is a small caveat here. I have discussed magnetoconvection in Section 5.1 and have pointed out that magnetic fields present in a convecting plasma tend to become concentrated in magnetic flux bundles. Will this not happen with the magnetic fields outside sunspots, since these fields are also located at the top of the sun's convection zone? In the 1970s careful solar observers such as Jan Stenflo and Robert Howard discovered that magnetic fields outside sunspots are indeed concentrated in small magnetic flux bundles, which are much smaller in size compared to sunspots. The typical radius of such a flux bundle is not more than a few hundred kilometres, whereas a large sunspot can be several thousand kilometres across. The typical magnetic field inside a small flux bundle is of order 0.1 tesla. The first magnetograms which measured the magnetic fields outside sunspots did not have sufficient power to show these small magnetic flux bundles separately. So what was measured by these low-power magnetograms was the value we would get if the magnetic fields of the flux bundles were smeared and distributed around over the solar surface. While interpreting the values of the magnetic field indicated by the bar in Figure 7.5, this fact should be kept in mind.

Since the magnetic field outside sunspots forming the bands shown in Figure 7.3 is essentially the poloidal magnetic field of the sun, the discussion in the last few pages should have given some idea of how the poloidal magnetic field at the sun's surface evolves with time. When tilted sunspot pairs decay, the magnetic fields of these pairs get spread around by turbulent diffusion to produce the poloidal magnetic field, and then this poloidal field is carried poleward by the meridional circulation. A group of scientists at the Naval Research Laboratory in Washington DC—Yi-Ming Wang, Anna Nash and Neil Sheeley— studied this process carefully in a series of landmark papers.[3] They took actual positions of sunspots from solar observations and then ran their numerical simulations to find out how the magnetic fields of these sunspots would get dispersed over the solar surface by turbulent diffusion and then carried poleward by the meridional circulation. They were able to explain the behaviour of the magnetic field outside sunspots as shown by the grey scale in Figure 7.5. Such analyses established that the poloidal magnetic field is indeed created at the sun's surface by the Babcock–Leighton mechanism.

The next step was to construct a full theoretical model of the solar dynamo in which the poloidal magnetic field is produced by the Babcock–Leighton mechanism rather than the α-effect (helical turbulence) as indicated in Figure 6.4. It was clear that the meridional circulation would play a very important role in this dynamo. In 1991 Yi-Ming Wang, Neil Sheeley and Anna Nash published a pioneering paper demonstrating that such a dynamo model really works.[4] It was a simple model in which a simple differential rotation was assumed rather than the differential rotation mapped by helioseismology as shown in Figure 3.9. As the simulations of magnetic buoyancy showed that the toroidal magnetic field of the sun is much stronger than what was previously assumed and the α-effect probably could not work inside the sun, some of us started getting interested in this new kind of dynamo model. I have already mentioned the principal difficulty facing us. If the poloidal magnetic field is produced by the Babcock–Leighton mechanism and we assume a strong differential rotation at the bottom of the convection zone, a straightforward application of the Parker–Yoshimura sign rule suggests that the dynamo wave should propagate poleward, producing sunspots at higher and higher latitudes with the progress of the solar cycle. This sign rule was based on an analysis without taking the meridional circulation into account. Could the meridional circulation turn things around and make the dynamo wave go in the correct direction? Since the 1991 paper of Wang, Sheeley and Nash had assumed a rather simple differential rotation, this paper did not provide a direct answer to this crucial question. The fact that the dynamo wave speed required to explain sunspot observations was almost precisely equal to the estimated equatorward speed of the meridional circulation at the bottom of the convection zone was very encouraging. But it remained to be demonstrated through a full mathematical analysis that the dynamo wave is really being carried by the meridional circulation this way by overruling the Parker–Yoshimura sign rule. We knew that the kind of dynamo model we were considering would work only if this could be convincingly demonstrated. Some of us decided to focus our attention on this critical question.

Before I describe our efforts towards answering this question, I would like to make a few remarks about one member of the group at the Naval Research Laboratory: Anna Nash. Unfortunately our research field remained a rather male-dominated field for a very long time. Nearly all the scientific results I have discussed in this book so far were obtained by men—the one exception being the Chinese-American woman Yuhong Fan, who made very important contributions on magnetic buoyancy. Anna, who was about the same age as me, was a pioneer woman scientist in our field. I had met her at a couple of conferences. She was a woman of stunning beauty and grace—the kind of woman who would attract attention even in a crowd. The epochal 1991 dynamo paper remains the last major work of this exceptionally talented woman. Shortly after publication of this paper, I received an e-mail from her colleague Yi-Ming Wang saying that Anna had taken her own life. I could hardly believe it. When I had met Anna only a few months earlier in a summer school in Crief in Scotland, she seemed so full of life. In tribute to Anna, I mention that now there are several leading women scientists in our field. Some of their contributions will be discussed in the remaining portions of the book. I take pride that I had the privilege of providing initial research guidance to some of these women scientists in getting them started in our field.

7.3 The Flux Transport Dynamo Model

One of the privileges of an academic job is sabbatical leave. Most universities and research institutes around the world allow an academic to work in a different place for a year after completing a few years (usually six) of service in the home institute. By the mid-1990s my sabbatical leave was due and I could spend a year away from the Indian Institute of Science. Luckily the Alexander von Humboldt Foundation offered me a Fellowship with which I could spend a year in Germany. I decided to spend the year in Kiepenheuer-Institut für Sonnenphysik in the beautiful city of Freiburg built around a magnificent medieval cathedral at the edge of the Black Forest.

One outstanding theoretical scientist in that institute was Manfred Schüssler (he left that institute a few years later)—a few years older than me and with research interests very similar to mine. Manfred and I would often discuss the solar dynamo problem. Bernard Durney of the National Solar Observatory in the United States was another person keenly interested at that time in the solar dynamo problem. I have already mentioned that many years ago Parker and I first heard about the new results of helioseismology from Bernard (Section 3.7). Towards the end of 1994, I received an e-mail from Bernard that he had some travel money left in a research grant. Since Manfred and I were both in Freiburg, he wanted to come there for a week to discuss with us. Both Manfred and I were excited and welcomed Bernard to Freiburg. During his week-long visit, almost every day Bernard, Manfred and I would have a brainstorming session lasting several hours. Bernard paid me probably the highest scientific compliment I have ever received in my life during one of those brainstorming sessions. After I suggested how we could formulate a certain aspect of our problem, Bernard suddenly said: 'Your way of looking at scientific problems is so similar to Parker's way of looking at scientific problems! Listening to you, I almost thought for a moment that I was listening to Parker.'

During these brainstorming sessions, we discussed the status of solar dynamo theory at that time, the unanswered questions to be addressed and the calculations we would have to do to address these questions. Many ideas which were hazy and nebulous in our heads became sharp and focussed as a result of these discussions. Since the flux rise simulations suggested a very strong toroidal field and helical turbulence cannot twist such strong fields, we all agreed that the Babcock–Leighton mechanism was our best bet for generation of the poloidal field. If such a dynamo model made the dynamo wave propagate poleward as suggested by the Parker–Yoshimura sign rule, then we would be in trouble. In order for a dynamo model of this type to work, we needed to check whether the equatorward meridional circulation at the bottom of the convection zone could turn this around and make the dynamo wave there go equatorward. That was the crucial question before us.

Mausumi Dikpati was working with me on her PhD thesis at that time. She had exceptional computational skills and had developed a computer code to study how the meridional circulation would carry the magnetic fields. I realized that in a few weeks I could extend Mausumi's code to a code for studying the solar dynamo problem with the Babcock–Leighton mechanism producing the poloidal field. I offered to do this. Only after the code got ready, would I be able to answer the question bothering us. Over the next few weeks following Bernard's visit, I devoted my entire energy to extending Mausumi's code to a full solar dynamo code. As soon as it was ready, I eagerly made runs to find an answer to our question. When I ran the code switching off the meridional circulation, the dynamo wave propagated poleward, as expected from the Parker–Yoshimura sign rule. Then I gave a run with the meridional circulation and waited for a few tense hours for the results to come. As soon as I made the plot, I knew that our idea worked. The dynamo wave was now moving towards the equator and would produce sunspots at increasingly lower latitudes with the progress of the solar cycle, in accordance with observations. Although the Parker–Yoshimura sign rule was derived for the situation without the meridional circulation, it was held to be an almost sacrosanct result for several years and many people expected that it would hold even when meridional circulation was present. My calculations at last showed that the meridional circulation could overturn this sign rule and could make the dynamo wave go in the correct direction, thereby paving the way for dynamo models in which the poloidal magnetic field is generated by the Babcock–Leighton mechanism. I immediately informed Manfred of this and sent an e-mail to Bernard.

Now we had the job of writing up our exciting results in the form of a paper. I asked Bernard if he would like to join us as an author of this paper. He declined, saying that his viewpoint differed from ours on some important (rather technical) matters and he would write a separate paper on his own. Since I had developed the dynamo code by extending the original computer code of Mausumi Dikpati, she was automatically included as an author. The paper jointly authored by me, Manfred Schüssler and Mausumi Dikpati appeared in 1995.[5] The separate paper by Bernard Durney also appeared in the same year.[6]

The type of solar dynamo model in which the meridional circulation plays an important part gradually came to be known as the *flux transport dynamo model*. I do not know who coined this horrible name. Many of us working on this topic strongly disliked this name and did not use it in our early publications. But somehow this name stuck and started being used by many solar physicists. Eventually those of us who were opposing this name yielded and at last started using this name. As I already mentioned, the feasibility of such a model was first demonstrated in the Wang, Sheeley and Nash 1991 paper using rather simplistic considerations. When the Choudhuri, Schüssler and Dikpati 1995 paper showed that one can get an equatorward propagating dynamo wave even in a model based on strong differential rotation at the bottom of the convection zone and the Babcock–Leighton mechanism at the surface (which would imply a poleward dynamo wave in the absence of meridional circulation), at last the flux transport dynamo model was accepted as a serious contender for explaining the sunspot cycle.

Over many pages of this book, I have discussed various ideas which have been combined together in the flux transport dynamo model. But it may be useful to put down the main ideas again in a brief form here. Figure 7.6 is a cartoon explaining the working of the flux transport dynamo. The strong differential rotation concentrated at the bottom of the convection zone would produce a strong toroidal magnetic field there out of any poloidal magnetic field present. The arrows indicate magnetic buoyancy which would make the strong toroidal field rise to the solar surface to produce sunspots. In a layer near the surface, the Babcock–Leighton mechanism produces the poloidal magnetic field from the decay of the sunspots which form out of the toroidal magnetic field. The contours with arrows indicate the all-important conveyor belt: the meridional circulation. This conveyor belt carries the poloidal field produced at the surface towards the pole, as seen in observations. Then this conveyor belt can take this poloidal magnetic field down to the bottom of the convection zone, where the differential rotation can act on this poloidal magnetic field to produce the toroidal magnetic field. Since the conveyor belt moves equatorward at the bottom of the convection zone, the toroidal magnetic field produced there moves equatorward with it, ensuring that sunspots appear at lower and lower latitudes with the sunspot cycle.

If you have understood the ideas encapsulated in Figure 7.6, then you have understood the basic ideas behind the flux transport dynamo

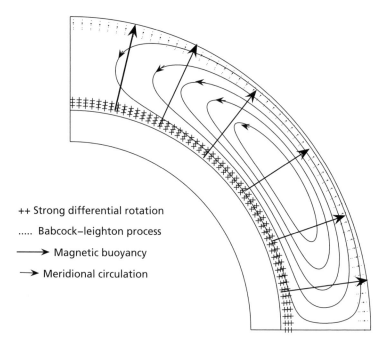

++ Strong differential rotation

..... Babcock–leighton process

⟶ Magnetic buoyancy

→ Meridional circulation

Figure 7.6 A cartoon explaining how the flux transport dynamo inside the sun works.

almost at the level at which an expert understands these basic ideas. What the expert has to do after this is to translate these basic ideas into mathematical equations and then solve them to explain different aspects of the solar cycle quantitatively. Certainly I cannot discuss the equations in a book of this nature. For readers curious to know what the equations look like, I show them in Figure 7.7. Although the notation used in these equations is quite standard in higher mathematical physics, professional training in physics over some years is required to be familiar with this notation and to acquire competence in handling such equations. The only way of proceeding with the equations shown in Figure 7.7 is to write a computer code to solve them. The code developed in Bangalore by me and my successive PhD students is named *Surya*, which is the Sanskrit word for the sun.

As I already discussed, one of the first technical issues which had to be sorted out was whether one can get an equatorward propagating dynamo wave by using the meridional circulation to overcome the

Magnetic field: $B(r, \theta)\mathbf{e}_\phi + \nabla \times \left[A(r, \theta)\mathbf{e}_\phi \right]$

Velocity field: $\mathbf{v} + s\,\Omega(r, \theta)\mathbf{e}_\phi$

$$\frac{\partial A}{\partial t} + \frac{1}{s}(\mathbf{v} \cdot \nabla)(sA) = \eta \left(\nabla^2 - \frac{1}{s^2} \right) A + \alpha B,$$

$$\frac{\partial B}{\partial t} + \frac{1}{r}\left[\frac{\partial}{\partial r}(rv_r B) + \frac{\partial}{\partial \theta}(v_\theta B) \right] = \eta \left(\nabla^2 - \frac{1}{s^2} \right) B$$

$$+ s\left(\mathbf{B}_p \cdot \nabla \right) \Omega + \frac{1}{r}\frac{d\eta}{dr}\frac{\partial}{\partial r}(r\,B),$$

where $s = r \sin \theta$ and $\mathbf{B}_p = \nabla \times \left[A(r, \theta)\mathbf{e}_\phi \right]$.

Figure 7.7 The basic equations which have to be solved to construct a model of the flux transport dynamo.

Parker–Yoshimura sign rule. I described how Choudhuri, Schüssler and Dikpati solved this problem in 1995. But there remained many other technical issues which had to be sorted out before the flux transport dynamo model could be taken as a satisfactory model of the sunspot cycle. Over the last few years, many groups around the world have written very important papers on these technical issues—giving us a deeper insight into the subject. But I cannot get into a discussion of these issues in a non-technical book like this. Many of my friends around the world who have written important papers on this subject over the last few years may be disappointed that I am not able to describe their works here. I assure them that this is not because I do not recognize the importance of their work. If I were writing a technical review meant for experts on this subject, I would certainly discuss how different groups around the world have addressed various technical issues connected with the flux transport dynamo model and put the model on a firmer footing. But doing that here would make this book too technical for my intended readership. In this book, I have described the historical developments leading to the formulation of the flux transport dynamo

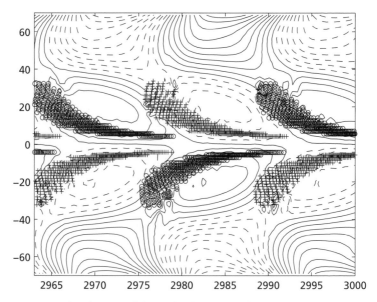

Figure 7.8 The theoretical butterfly diagram including both sunspots and the magnetic field outside sunspots. Taken from Chatterjee, Nandy and Choudhuri 2004.

model and elucidated the main ideas behind this model. One cannot go much beyond that in a non-technical book like this.

To give an idea of the level of sophistication that the flux transport dynamo model reached in a few years, Figure 7.8 reproduces a figure from a 2004 paper from our group—written with two of my PhD students: Piyali Chatterjee and Dibyendu Nandy.[7] This is a theoretical figure obtained from our dynamo model to be compared with Figure 7.5 based on observational data. The butterfly diagrams in the two figures can be compared easily. The contour lines in Figure 7.8 indicate how the magnetic field on the solar surface outside sunspots evolves in latitude and time in our theoretical model. These contour lines have to be compared with the grey-scale background in Figure 7.5. Given the fact that this was about the first serious theoretical effort to model the observational data shown in Figure 7.5, I hope that most of the readers will agree that the match between the observational Figure 7.5 and the theoretical Figure 7.8 is not too bad.

With the publication of papers like the Chatterjee, Nandy and Choudhuri 2004 paper (from which Figure 7.8 is taken), the flux transport dynamo model reached a stage where it could be used to explain most of the periodic features of the sunspot cycle. But the sunspot cycle is not completely periodic, as a look at Figure 1.5 would convince anybody. What causes these irregularities, including such extremes as the Maunder minimum? If we understand the causes behind the irregularities of the cycle, can we predict how strong the upcoming cycle will be? Once the flux transport dynamo model succeeded in reproducing the various periodic features of the sunspot cycle, several groups around the world (including our group) started using this model to understand the irregularities of the sunspot cycle. I would say that this is one of the main thrusts of research in this field at the present time. Research on this topic of irregularities of the sunspot cycle is still of a rather tentative nature and has not yet achieved a level of general agreement among the various groups around the world. The ideas of one group are often disputed and contested by another group. I shall give a brief introduction to this fascinating subject in Chapter 9. But I want to do something else before that. I started this book by telling you how the earth gets affected by solar disturbances. I promised that I shall give an explanation of the long chain of causes-and-effects which connect phenomena on the sun with phenomena on the earth. Now that I have discussed dynamo theory which tells us why the sun has magnetic fields and why there is a sunspot cycle, it is time to turn to the question of why the magnetic field of the sun causes such explosions like solar flares and why they affect the earth so far away from the sun. We need to understand the physics of the sun's corona to address these questions. This is the topic which will be taken up in the next chapter. Only after that, in Chapter 9 we shall come to a discussion of the irregularities of the sunspot cycle, which is a hot topic of current research.

Solar dynamo theory has come a long way since Parker's monumental 1955 paper on this subject. Although that paper has been like a guiding beacon for all subsequent developments in this field, we now think that some of the ideas of that paper are not applicable to the sun. Parker's original idea of how the toroidal magnetic field is twisted by helical turbulence to produce the poloidal magnetic field would not work if the toroidal magnetic field is as strong as we now believe it to be. The dynamo wave derived by Parker based on his idea—although a tremendously important historical step in the development of this subject—is no longer believed to provide the correct model of the sunspot cycle. However, Parker's idea of helical turbulence twisting one component of the magnetic field to produce another component seems to work in other astronomical systems. For example, it is believed that the magnetic field of the earth is produced and maintained more or less the way Parker envisaged. The magnetic field in the interior of the earth is not believed to be extremely strong in any region. So helical turbulence can twist this magnetic field exactly in the way Parker suggested.

Ever since I had been involved in developing the flux transport dynamo, I wondered how Gene Parker would take our results. While visiting the United States in 2000, I made a stop in Chicago to give a seminar on my work. Gene had retired several years earlier and came to the University only occasionally. He told me that he would come on the day of my seminar. He also invited me to his house for dinner in the evening after my seminar. As it happened, I was meeting Gene exactly 11 years after our previous meeting: a full sunspot cycle later! As Gene came in, I saw that he was walking more slowly than before and he had more grey hair. But otherwise he was his old cheerful self. After we exchanged pleasantries for a few minutes, Gene asked me to explain the dynamo model I was working on. Gene was not keeping track of the literature in our field as closely as he did in earlier times. So I rose to the blackboard to explain the flux transport dynamo model. It was one of those poignant moments which rarely come in a scientist's life. Seated before me was the man whom I revered more than any other living astrophysicist. By a strange twist of fate, it had befallen me to discover that some ideas in one of his most celebrated papers did not work for the sun and now I had the job of telling him that. Even when I was his PhD student and had to explain my work to him, I had probably never been as nervous as I was that day. As I started describing the developments leading to the flux transport dynamo model, he listened with intense concentration, repeatedly interrupting me with questions. After I finished, he looked up with dreamy eyes, as if he was looking at something far beyond me. Then he said slowly: 'You know, Arnab, I wrote that dynamo paper of mine many years ago when very little was known about the conditions in the interior of the sun's convection zone. Now we know so much more and you guys have developed a new model in response to that new knowledge. That is the right way to go. That is how science progresses.'

7.4 A Critique of the Flux Transport Dynamo

We have seen that the flux transport dynamo model is able to explain various regular periodic aspects of the sunspot cycle. We shall see in Chapter 9 that this model has been remarkably successful in explaining the irregular aspects of the sunspot cycle as well. Does this mean that the flux transport dynamo model is the correct theoretical model of the sunspot cycle?

I would not be distorting the truth if I say that the majority of solar physicists regard the development of the flux transport dynamo as a breakthrough in our field. However, there is a small number of quite eminent solar physicists who are sceptical about the flux transport dynamo model. Although their number is small, I feel that I ought to give my view on this matter. Let me give the analogy of a clock. Suppose somebody looks at the motions of the hour hand, the minute hand and the second hand, and then tries to make a theoretical model of the interior mechanism. If he is able to come up with a mechanism that would exactly give rise to the observed motions of the clock hands, can we be sure that this is the actual mechanism inside the clock? Unless we can open the clock and compare the theoretical model with what is inside, we can never be completely sure.

In the flux transport dynamo model, we have been able to come up with a possible mechanism within the sun's convection zone that explains various observations of the sunspot cycle. But is our theoretically proposed mechanism really what is happening inside the sun? Let us look at the basic assumptions in the theoretical model. One assumption is that the toroidal magnetic field is produced by the differential rotation measured by helioseismology as shown in Figure 3.9. Even the staunchest critics do not question this assumption. The next assumption is that the poloidal magnetic field is produced by the Babcock–Leighton mechanism at the surface from the decay of sunspot pairs. We find clear evidence of this process from observations of the solar surface. It is true that this is a complex process which is incorporated in the present-day theoretical models in a simplistic and crude way. We expect significant improvements in modelling magnetic buoyancy and the Babcock–Leighton process in the near future. Already a first step has been taken in this direction in an important recent paper by Anthony Yeates and Andres Muñoz-Jaramillo.[8] Although our handling of these things at present may not be fully satisfactory, it does not seem likely that we are completely off the track. Another important parameter in the theoretical model is the value of turbulent diffusion. As we shall discuss in Chapter 9, there is some uncertainty in its value, although the irregularities of the sunspot cycle put important constraints on the value of turbulent diffusion. Since flux transport dynamo models work for a range of values of turbulent diffusion, our lack of knowledge of its exact value can hardly be offered as a reason for discarding the flux transport dynamo model altogether. The one other remaining assumption is the assumed nature of the meridional circulation in our theoretical models. We need to have an equatorward meridional circulation at the bottom of the convection zone to get around the Parker–Yoshimura sign rule. We have observations of the meridional circulation near the surface and the assumed meridional circulation in the theoretical models is made to match these surface observations. But, to this day, we do not have a direct proof that the meridional circulation at the bottom of the convection zone is really like what the theoretical models assume. To my mind, this is the only crucial assumption of the flux transport dynamo model which is neither proved nor disproved at the present time. Since the flux transport dynamo model has been so successful in explaining so many aspects of the sunspot cycle, most of us consider it extremely unlikely that a fundamentally wrong theoretical model would have all these successes merely by accident and believe that the meridional circulation at the bottom of the convection zone must be like what is assumed in this model. But, to some extent, this is clearly a matter of taste. Someone can hold the opposite viewpoint: 'There is no proof that the meridional circulation at the bottom of the convection zone is like what is assumed in flux transport dynamo models. In my opinion, the meridional circulation is not like that and the flux transport dynamo model is not correct.'

I should mention that very recently (within the last couple of years) some three groups have claimed that they find evidence for the return flow of the meridional circulation in the middle of the convection zone rather than at the bottom. Their results are based on the analysis of data in which it becomes increasingly difficult to distinguish the signal from noise as one goes deeper into the convection zone. Also, the results of these different groups so far do not agree with each other in detail. So I would say that the last word on this subject has not been said yet. With my students Gopal Hazra and Bidya Karak, I have recently shown that some of the attractive features of the flux transport dynamo can be retained even if the meridional circulation turns out to be more complicated than what is usually assumed in theoretical models.[9] Since this is a rather technical subject, I shall not discuss it in any more detail.

Let us ask the hypothetical question: what will happen if the flux transport dynamo theory is really proved wrong after all? Even the harshest critics of the flux transport dynamo agree that this theory has a phenomenal success in explaining many aspects of the sunspot cycle (which they consider pure accident) and that there is no other rival theoretical model which can explain these things satisfactorily. So, if the flux transport dynamo fails, we have to conclude that we do not understand at all how the solar dynamo works and we have to start again from the beginning.

Let me now say a few words on how the flux transport dynamo is received by the solar physics community at large. Whether other scientists in a field consider a paper to be important is often judged by the number of citations to the paper by other scientists. I counted that, since 1995 when the Choudhuri, Schüssler and Dikpati paper on the flux transport dynamo appeared, there have been at least a dozen papers on the flux transport dynamo by different groups which have received more than 100 citations to date. During the same period, I am aware of only three papers dealing with alternative models of the solar dynamo which have exceeded 100 citations. In many international conferences which cover the sunspot cycle, there is often an invited review on the solar dynamo and usually somebody working on the flux transport dynamo is invited to give the review. I have been invited to give such reviews on several occasions in the last few years. Some of my former students such as Mausumi Dikpati, Dibyendu Nandy or Jie Jiang would also often be asked to give such a review. I shall discuss in Chapter 9 the differences I have with Mausumi. But both she and I agree on the fundamental premise that the flux transport dynamo is a very promising model for the sunspot cycle. Another distinguished person in this field who is regularly asked to give invited reviews is Paul Charbonneau, who has made tremendously important contributions to the flux transport dynamo in the last few years. The fact that one of us working on the flux transport dynamo would be invited to review the solar dynamo at virtually any important international conference covering this subject shows that the solar physics community regards the flux transport dynamo model to be a model which should be taken very seriously at the present time.

We shall discuss in Chapter 9 how the flux transport dynamo model had some remarkable successes in explaining the irregularities of sunspot cycles and in predicting future cycles. When some of our papers on these subjects were published, Dr. R. Ramachandran, a leading science journalist of India, decided to write a fairly long article on this subject for the magazine *Frontline*.[10] He contacted two or three solar physicists for comment. As it happened, the persons whom Ramachandran contacted belonged to the small group of solar physicists highly critical of the flux transport dynamo. Ramachandran told me that he was surprised by the strong reactions he encountered, since he did not expect such reactions about a theory which, on the face of it, seemed very successful in explaining so many aspects of the sunspot cycle. Professor Steven Tobias of Leeds University, who wrote a glowing review of my book *The Physics of Fluids and Plasmas* several years ago, commented: 'I would warn you against confusing a successful prediction with a correct model. These models are fraught with difficulties and have a number of ad hoc assumptions that cannot be justified.' Professor Horamzad Antia of Tata Institute of Fundamental Research in Mumbai was more blunt: 'Dynamo theory is a major unsolved problem, and I don't think we are anywhere close to a breakthrough in this area.' When this article appeared in *Frontline*, some friends asked for my response to Professor Antia's remark. I give here the same response which I gave to my friends at that time: 'No comment.'

8

A Journey from the Sun to the Earth

8.1 Strange Connections

I began this book by mentioning how solar explosions cause havoc on the earth. I also said that, only towards the end of the book, will we be able to discuss how this happens. In the previous few chapters, I have discussed how such remarkable concentrations of magnetic flux as sunspots arise and why they have a cycle of 11 years. Now at last we are ready to address the questions why spectacular explosions like solar flares sometimes occur above sunspots and what are the exact mechanisms by which these explosions cause various dramatic effects on the earth.

Often a new field of research starts with the discovery of certain correlations: the discovery that two apparently completely different phenomena occur simultaneously, indicating some connection between them. That is how this field started. In 1741 the Swedish astronomer Anders Celsius, who is now mainly remembered for introducing the now universally used Celsius scale of temperature, and his assistant Olof Hiorter noticed that polar aurorae were associated with geomagnetic storms. We have already discussed geomagnetic storms (Section 1.1), which are sudden changes in the earth's magnetic field. Celsius and Hiorter discovered that almost always there are geomagnetic storms when polar aurorae appear. Although nobody at that time knew what caused polar aurorae or what caused geomagnetic storms, it seemed that there is some kind of a mysterious connection between them.

The first person to realize that both polar aurorae and geomagnetic storms have something to do with the sun was Edward Sabine in Victorian England. He realized that a large amount of good data about the earth's magnetic field was needed to solve the mystery of geomagnetic storms. He carried out a campaign to establish several magnetic observing stations around the vast British Empire of his time. As he kept accumulating more data, he noticed a curious pattern. In some years, lots of geomagnetic storms occurred and in other years, very few. The frequency of occurrence of geomagnetic storms seemed to go up and down with a periodicity of 11 years.

In 1844 Heinrich Schwabe had reported the discovery of the 11-year sunspot cycle. At first this discovery attracted very little attention. One person who was intrigued by this discovery was Alexander von Humboldt, one of the towering figures of German science at that time. Humboldt had a plan to put all the knowledge about the natural world available at that time in a series of volumes named *Kosmos*. Schwabe's discovery of the sunspot cycle was reported promptly in the pages of *Kosmos*. As it happened, Edward Sabine's wife Elizabeth was involved in the mammoth project of translating *Kosmos* from German to English. Edward Sabine used to go over the pages of his wife's manuscript. That is how he came to know about Schwabe's discovery. He immediately compared his own geomagnetic storms data with Schwabe's data of sunspots. He was completely astounded. More geomagnetic storms occurred precisely when there were more sunspots. The two cycles were one and the same. Sabine reported this path-breaking discovery in 1852.[1] It seemed that sunspots were the cause of both the aurorae and geomagnetic storms.

The 1859 discovery of the solar flare by Richard Carrington gave the first clue as to how such a connection may arise. As I already mentioned in Section 1.1, Carrington also found that a violent geomagnetic storm followed 18 hours after the solar flare. Reports of unusually brilliant auroral sightings also came from various places around the world—even from places where aurorae are not usually seen. The intriguing idea that almost suggested itself was that sunspots often give rise to solar flares occurring above them and then these solar flares cause both the aurorae and geomagnetic storms. In such a scenario, occurrences of flares, aurorae and geomagnetic storms would all vary in tandem with the 11-year sunspot cycle. I discussed that Lord Kelvin, the most important physicist in Britain at that time, ridiculed this idea in 1892, since he thought that this idea was against all the laws of physics known at that time. But observational support for this idea was so overwhelming by the closing years of the nineteenth century that it became one of the scientific challenges of the time to probe the laws of physics deeper and to explore how such incredible connections may arise. Why do sunspots give rise to such explosions as solar flares? How do flares produce the aurorae and geomagnetic storms in the earth? It took scientists several decades to come up with answers to these questions.

Since polar aurorae are seen regularly in Scandinavian countries, scientists of Scandinavian countries have always had a special interest in them. In the early years of the twentieth century, two Norwegian scientists—Kristian Birkeland, an experimenter, and Carl Størmer, a theorist—came up with an explanation for aurorae. Around that time (in 1897) J. J. Thomson had discovered the electron from his studies of the discharge tube. In a discharge tube, an electric current is made to flow through a very rarefied gas. The fast-moving electrons in this current can ionize some atoms of the rarefied gas, knocking off electrons from these atoms. These ionized atoms then try to attract nearby electrons. When an electron is captured by the ionized atom to make it neutral again, some energy is lost in the form of a photon. Photons emitted in this way cause a glow in the discharge tube. Birkeland noted that the glow emitted from a discharge tube was rather similar to the glow of the aurorae, suggesting a common physical origin. He carried out an experiment in which the usual plane cathode of the discharge tube was replaced by a metallic sphere with a strong magnetic field, resembling the geomagnetic field as indicated in Figure 2.10(b). Birkeland found that the glow was concentrated near the poles of this magnetic sphere, exactly as is the case with the aurorae which occur above the earth's magnetic poles.[2] Birkeland's colleague Størmer carried out mathematical calculations to understand how this happens. The force acting on a moving charged particle is called the Lorentz force—named after Hendrik Lorentz, who figured out the mathematical expression of this force. A standard topic that is discussed in high school physics textbooks is finding out how a charged particle will move in a uniform magnetic field. It can be shown by quite elementary mathematical reasoning that the Lorentz force makes the charged particle move around a magnetic field line in a helical path. Similar calculations for charged particles moving in the magnetic field of a spherical magnet, as shown in Figure 2.10(b), are much more complicated. Størmer carried out these calculations and showed that the charged particles still go around the magnetic field lines, which divert the charged particles towards the poles, as indicated in Figure 8.1.[3]

From Birkeland's experiment and Størmer's mathematical calculations, an explanation for the aurora emerged. Suppose a solar flare causes the discharge of an electric current. The charged particles of this current would be diverted by the earth's magnetic field towards the geomagnetic poles. When these charged particles encounter the thin

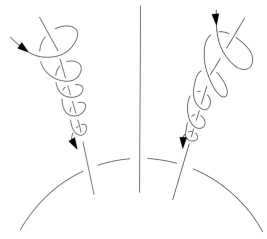

Figure 8.1 A sketch indicating paths of charged particles moving in the magnetic field of a spherical magnet. In reality, there is an additional complication (the charged particles have a tendency to revolve around the magnetic axis), which is not shown to keep the figure simple.

air in the upper atmosphere, some atoms of the air get ionized by them and the glow is produced exactly the same way as in the discharge tube. The only problem with such a model is the idea that a solar flare produces an electric current which flows from the sun to the earth. Several leading physicists of that period regarded this as a completely absurd idea and refused to accept this model of the aurora. However, over the next few decades, various kinds of supporting evidence kept coming for the Birkeland–Størmer model of the aurora. We now regard this to be a basically correct model.

We then have to address the question of how solar flares give rise to currents. We now have evidence that solar flares can accelerate a handful of charged particles to very high speeds. When we discuss the solar wind later in this chapter, we shall point out how these fast moving charged particles reach the earth to produce aurorae starting from their site of acceleration: the solar flares. Let me now make a few remarks about the acceleration of charged particles. A solar flare is not some unusual astronomical event which causes the acceleration of some charged particles to high speeds. For nearly a century now, scientists have been aware that the earth is continuously bombarded by charged particles moving at speeds close to the speed of light coming from all the directions.

Let us begin at the beginning. In the early years of the twentieth century, physicists were puzzled to find that air in the atmosphere is always slightly ionized. The question was to find the agent causing this ionization. A clue to answering this question came from the discovery that this ionization increases with height in the atmosphere. The person who made this discovery in a series of daring experiments was Victor Hess, who was to win the 1936 Nobel Prize for this discovery. That was a time when scientists would often carry out risky experiments at great personal danger to themselves. Even by the standards of that time, what Hess did was really daring. He took instruments for measuring ionization in balloon flights rising to altitudes as high as 5400 metres. In 1912 Hess announced his discovery that the ionization increased with altitude, clearly indicating that the ionization was caused by something coming from outside the earth.[4] This something coming from outer space was named *cosmic rays*, in the belief that it would be some kind of radiation. Although it was later found that this something consisted of charged particles moving at speeds close to the speed of light rather than radiation, the name cosmic rays persisted. The cosmic ray particles move so fast that they are not much affected by the earth's magnetic field and seem to come from all directions (although there is a slight variation in cosmic ray intensity with the earth's latitude), unlike the less energetic charged particles coming from solar flares which are channelized to the earth's geomagnetic poles. The most commonly occurring particles in cosmic rays are electrons, protons and light nuclei such as the nuclei of helium. Many important properties of elementary particles can be studied by making them move at very high speeds and collide with each other. That is why physicists have built huge machines to accelerate elementary particles to very high speeds—the largest of such accelerators being in CERN Geneva. In the first half of the twentieth century, before such large particle accelerators were built, cosmic rays were the only source of highly accelerated particles for physicists to study. Some elementary particles such as the positron (the anti-particle of the electron) and the mesons were first discovered in cosmic rays.

I decided to mention a few basic facts about cosmic rays because we shall have to encounter them while discussing some other topics in the remainder of the book. But I want to keep away from the crucial astrophysical question of how a handful of charged particles get accelerated to such high speeds. It is known as the *particle acceleration problem* in astrophysics. This is a vast subject which can be the topic for another full book and is completely outside the scope of the present book. I can only make a few superficial comments without getting into details. In this book, the central theme is the dynamo process which generates magnetic fields in different astronomical systems like the sun. It is believed that the magnetic fields of some astronomical bodies are responsible for the particle acceleration. How this happens is somewhat subtle. One famous result of electromagnetic theory which every student has to learn in high school is that a static magnetic field cannot change the energy of a charged particle. In other words, when a charged particle moves in a static magnetic field, the Lorentz force of the magnetic field deflects the charged particle to move in a spiral path, but cannot change its speed. Something more complicated has to happen for charged particles to be accelerated by magnetic fields. Some ideas of how this may happen were given in a famous 1949 paper by Enrico Fermi, one of the most versatile physicists of the twentieth century.[5] Many astrophysicists afterwards elaborated on Fermi's ideas and provided a scenario of how charged particles get accelerated to very high speeds in some astrophysical situations.

Although I shall not say anything more on how some charged particles get accelerated to high speeds in a solar flare, I need to discuss what causes such explosions as solar flares and how they eventually cause geomagnetic storms. To address these questions, we first have to understand the mysterious corona surrounding the sun.

8.2 The Beautiful Corona and a Wind from it

Those of you who had been lucky to witness a total solar eclipse would have seen the beautiful corona surrounding the darkened disc of the sun. It is one of the most awesome sights in the natural world. Figure 8.2 shows a view of the corona photographed during a total solar eclipse. Since the corona is much fainter than the sun's disc, it becomes visible only when the light from the sun's disc is obstructed. Until a few decades ago, astronomers had to wait for a total solar eclipse to see the corona. Then in the 1930s the French astronomer Bernard Lyot invented the coronagraph, in which an artificial eclipse is created inside a telescope by blocking the sun's disc. However, there is so much

Figure 8.2 The solar corona photographed during a total solar eclipse on 26 July 2009. Credit: Koen van Gorp.

scattered light at sea level due to the scattering of sunlight by atmospheric particles that one cannot see the corona from sea level even after blocking the sun's disc. Only if one takes a coronagraph to a high mountain, can the corona be seen without an eclipse. Nowadays the solar corona is monitored continuously from coronagraphs carried on spacecraft. Since there is no problem of scattered light in space, one can get a pristine view of the corona from space.

Even in the nineteenth century, astronomers were making expeditions to the sites of the total solar eclipse, one aim being to study the corona. I have already mentioned Eddington's eclipse expedition in 1919 to study the bending of light, which proved Einstein's general theory of relativity (Section 3.4). Writing in the middle of the twentieth century, the eminent astronomer van de Hulst gave the following account of the hazards of an eclipse expedition:

> A year's preparation at home, a long journey, and a month's preparation on the eclipse site are required. Clouds may make all this work useless. If the sky is clear, some member of the expedition may contract malaria or break his arm. A motor may not work after many flawless rehearsals, or in the excitement of the great moment the person in charge may forget to take the cover off the telescope. All these events have happened.[6]

A particularly challenging job was to take the spectrum of the corona during the few minutes of the total solar eclipse. Although the faint light from the corona is primarily sunlight scattered by the electrons of the corona, the corona itself also emits some light. During a total eclipse seen from southern India in 1868, Pierre Janssen succeeded in taking a spectrum of the corona and saw a bright yellow line. This line was also seen by Norman Lockyer, the founder of the famous journal *Nature*. Lockyer realized that this was a spectral line never seen in the laboratory until that time and that it must correspond to an unknown element. He named it helium from the Greek word 'helios' for the sun. Helium was discovered on earth about ten years later by William Ramsay.

Even at the beginning of the twentieth century, some of the lines seen in the spectrum of the corona still remained unidentified. It was guessed that these lines might be due to some still unknown element *coronium*. However, by that time, Mendeleev's periodic table was fairly filled up and hardly any vacancies remained for an unknown element. Finally the puzzle was solved by Bengt Edlén in 1943.[7] He realized that some spectral lines are signatures of iron atoms which have lost as many as thirteen electrons! We have discussed Saha's famous theory of thermal ionization (Section 4.2). According to this theory, as the temperature of a gas increases, more and more electrons get knocked out of the atoms in the gas. Edlén estimated that a vapour of iron has to be heated to a temperature of about a million degrees for iron atoms to lose so many electrons and concluded that the solar corona should have such a high temperature. Even though the sun's surface has a temperature of less than 6000 degrees, this surface is surrounded by an atmosphere—the corona—having a temperature in excess of a million degrees.

This conclusion seemed to be so incredible that initially there were very few physicists ready to accept it. In fact, until the middle of the nineteenth century, some astronomers had doubted whether the corona represented the atmosphere of the sun at all. Many astronomers thought that the corona was the moon's atmosphere illuminated by the sun from behind during the total eclipse. Only when the spectrum of the corona was taken and was found to be like the spectrum of a hot gas, did astronomers accept that the corona must be the hot atmosphere of the sun. The moon was unlikely to have such a hot atmosphere. However, a temperature of a million degrees seemed too incredible. This also meant that, although the sun was the ultimate source of heat, the temperature was higher in regions farther away from the sun—from the surface to the corona. This seemed to make no sense. Suppose you are camping out on a cold night. You would like to be near the campfire to be warm. You would certainly be very surprised if you found that it felt warmer as you moved farther away from the campfire. But something like this seems to be happening around the sun.

One of the most important and profoundest laws of physics is the second law of thermodynamics. C. P. Snow, in his famous book *The Two Cultures*, proposed that a familiarity with this law can be taken as a test for one's scientific culture, just as a familiarity with *Hamlet* can be taken as a test for one's literary culture. One version of this law roughly states that heat flows from regions of higher temperature to regions of lower temperature. If the corona is really so hot, then apparently the opposite of this happens. The heat of the sun flows out from the surface, which is at a lower temperature, towards the corona, which is at a much higher temperature. Is this a violation of the second law of thermodynamics? Is this consistent with the basic principles of physics?

Those of you who have studied thermodynamics systematically in a college course would know about the concept of thermodynamic equilibrium. When different parts of a system interact with each other, they are expected to reach thermodynamic equilibrium. If a system is far from thermodynamic equilibrium, one has to be careful in applying the principles of thermodynamics to such a system. The solar corona is a system very far from thermodynamic equilibrium. The corona is composed of low-density hot gas, through which the photons of light emitted from the solar surface travel outward. Only occasionally may a photon be scattered by an electron in the corona, making the corona visible when the glare of the sun's surface is hidden. The sun's surface appears as a sharp surface to us because the corona is almost transparent to the photons. In other words, the hot gas in the corona and the stream of photons flowing out from the sun interact with each other very little. They are nowhere near reaching thermodynamic equilibrium. The photons flowing outward do not know that there is a hot gas existing there. So we cannot apply the second law of thermodynamics in the usual way to this situation.

Even if we accept that the existence of the very hot corona does not contradict the principles of thermodynamics, we still do need an explanation for it. One of the first things to note is that, because of the very low density of the corona, the quantity of heat contained within the corona is not high in spite of its high temperature. Suppose you somehow managed to extract all the heat out of one cubic kilometre of the corona and put it in your cup of coffee. The temperature of the coffee would go up by only a few degrees! That is why a space vehicle sent to the outer regions of the corona (not too close to the sun) is not expected to be roasted by the heat of the corona. The high temperature of the corona simply means that all the particles in the corona are moving around at very high random speeds. Not only does the corona not have that much heat, the corona is also radiating away its heat all the time. A simple estimate shows that, if a continuous supply of heat to the corona did not exist, then the corona would radiate away all its heat in a few hours and would disappear. Such estimates demonstrate that the corona is a physical system in a highly precarious state and needs a continuous supply of heat in order to exist, like a vampire needing a continuous supply of blood from his victims. However, the heat requirement of the corona is not much because of its low density. If only a mere 10^{-5} part of the heat coming out of the sun can be dumped into the corona, that will be sufficient for the corona to exist. The heat from the sun passes through the corona in the form of radiation (or a stream of photons, if you prefer to use a more microscopic description). If only 10^{-5} part of this radiation gets absorbed in the hot gas of the corona, that may provide an explanation for the heat source of the corona. But the corona is so transparent that even this small amount of radiation does not get absorbed in it. So we need some alternative explanation for the heat supply to the corona.

In the next section, we shall come to the question of what heats the corona. Right now let me discuss another completely independent argument that the sun's atmosphere must indeed be extremely hot. Suppose we consider the atmosphere of our earth. It thins out considerably if we go up 10 kilometres above the ground. How the density of air falls with height can be calculated from the hydrostatic equilibrium equation (discussed in Appendix B), which is a cornerstone in the theory of stellar structure as well. The earth's gravity is pulling the air in our atmosphere and is not allowing the atmosphere to extend very far. Now imagine what would happen if the temperature of the earth's atmosphere were to increase by 500 degrees. One minor consequence of such an increase in temperature would be that all life on the earth would perish. But let us focus our attention only on the atmosphere.

All the molecules in air would move around with much higher speeds if the temperature of the atmosphere were raised like that. As a result, the earth's gravity would find it harder to hold on to the atmosphere and the atmosphere would extend much further. One can easily show from the hydrostatic equilibrium equation that an atmosphere extends more against a gravitational field when its temperature is more. Now, we know that the extension of the corona compared with the sun's radius is much, much more than the extension of the earth's atmosphere compared with the earth's radius. This suggests that the corona of the sun must be very hot. Such an argument could, in principle, be given even in the middle of the nineteenth century. Probably nobody gave such an argument at that time because most scientists thought of the corona as some kind of ghostly radiation and did not realize that it is made up of gas to which the hydrostatic equilibrium equation can be applied.

One of the first scientists to apply the hydrostatic equilibrium equation to the sun's corona was the distinguished geophysicist Sydney Chapman. Then in 1958 Eugene Parker carried out a famous analysis of the problem. At that time, nothing was known about the mechanism by which heat was supplied to the corona. Parker realized that it should be possible to build a model of the corona even without knowledge of its heat source. Consider the case of a metal rod of which one end is kept in a furnace. Suppose we want to find out how the temperature of the rod falls off as we move away from the furnace. If we just know the temperature produced by the furnace, we can analyse this problem without knowing whether the furnace is heated by charcoal, gas or oil. In exactly the same way, if we just assume that some unknown mechanism is depositing heat in the lower layers of the corona to raise its temperature to a million degrees, we can model the outer regions of the corona. When Parker constructed a hydrostatic model of the outer corona, he found something totally baffling. He expected both the temperature and the pressure of the coronal gas to fall to zero far away from the sun. He could make the temperature zero at infinity in his model. But, very surprisingly, he found that his mathematical equations would not allow the pressure to go to zero at infinity. The hydrostatic model yielded a finite non-zero pressure at infinity. This was extremely puzzling. This meant that something at a large distance from the sun has to apply this non-zero pressure to keep the corona in hydrostatic equilibrium. There was a time when astronomers believed that planets and stars were embedded in crystal spheres turning around their axes. If there was such a crystal sphere surrounding the sun at some distance from it, then that crystal sphere could presumably provide the necessary non-zero pressure to keep the solar corona in hydrostatic equilibrium. But this idea of crystal spheres was dismissed after Newton provided an explanation of the solar system from his theory of universal gravitation. Could there really be such crystal spheres?

Parker drew a conclusion which seemed at that time to be only marginally less radical than proposing a crystal sphere around the sun. If there was nothing to provide this non-zero pressure and keep the corona in hydrostatic equilibrium, Parker concluded that the corona would keep on expanding at a steady rate and would not be in hydrostatic equilibrium. The equations of hydrostatic equilibrium give inconsistent results simply because the corona is not in hydrostatic equilibrium and these equations are not applicable. Parker used the full equations of fluid mechanics and found that the outer part of the corona should keep on expanding steadily in the form of a plasma outflow from the sun. Parker named this the *solar wind*. It is basically the high temperature of the corona which makes it impossible for the sun's gravitational field to keep the corona confined and causes the solar wind to flow out. From the mathematical equations, Parker was able to calculate the density and the speed of the solar wind near the earth's orbit on the basis of some reasonable assumptions. He wrote up his results in the form of a paper and submitted it to *The Astrophysical Journal*. Two referees surmised that the paper was completely absurd because it went against the conventional wisdom of the time. They recommended that the paper should be rejected. After receiving these strongly negative referee reports, Subrahmanyan Chandrasekhar, the editor of *The Astrophysical Journal*, decided to read the paper himself. He could not find any flaws in the mathematical calculations or in the arguments, even though the final conclusion was so incredible. I wonder whether the referees' reactions reminded Chandrasekhar of Eddington's reaction to his own work on the mass limit of white dwarf stars many years ago. Chandrasekhar decided to overrule the referees and published the paper in *The Astrophysical Journal*.[8] The paper was destined to be probably the most influential theoretical paper of all time in the history of space science. It completely altered our conception of the space around our planet earth. It suggested that the space surrounding the earth is not empty, but rather the earth is immersed in a continuous plasma outflow from the sun. I may mention that all the mathematical calculations in Parker's 1958 paper proposing the solar wind are extremely simple. In fact, they are so simple that a superficial perusal of the paper may give one the deceptive impression that even an average scientist with an average competence in fluid mechanics could have done these calculations. But nobody except Parker realized that one could look at the problem in the way he did. It happens not too infrequently that truly original works of science are also works of utmost simplicity.

There were certain other important observations which could be explained with the idea of the solar wind. For example, tails of comets always extend away from the sun. It was initially thought that the pressure of sunlight was responsible for this. However, careful analysis shows that the pressure of sunlight is not quite enough to blow away the tails of comets. In the 1950s Ludwig Biermann proposed that only a stream of particles flowing from the sun would be able to turn the comet tails away from the sun. The solar wind appears to be just the right thing to do this job. Still the idea of the solar wind was not generally accepted until space vehicles provided indisputable proof of its existence. As I shall discuss later in this chapter, the solar wind is blocked by the earth's magnetic field. So one has to go above what is called the earth's *magnetosphere* to detect the solar wind directly. By a remarkable historical coincidence, the Space Age dawned at about the same time when the solar wind was predicted. The first artificial satellite Sputnik I was launched into orbit on 4 October 1957—almost precisely at the time when Parker was developing his theory of the solar wind. Within four or five years, measurements from space vehicles proved beyond any doubt that the solar wind is really there. The Soviet space vehicles Lunik I and Lunik II were the first to carry out such measurements. By the mid-1960s, the speed and the density of the solar wind near the earth's orbit could be measured and were found to roughly agree with what was predicted by Parker's model, triumphantly validating this model.

The solar wind is certainly taking away mass from the sun continuously. You may wonder whether one day the sun will disappear, losing all of its mass to the solar wind. You need not worry about such a calamity. Since the density of the expanding corona is so low, the sun is losing its mass at a very slow rate. During the 4.5 billion years since its birth, the sun is estimated to have lost only about 10^{-4} of its mass. Presumably the sun will become a white dwarf much before it loses a substantial amount of mass through the solar wind. The lower corona where the density is higher is in hydrostatic equilibrium to a very good approximation. Only in the higher corona, does the expansion start, initially at a low speed. However, at a substantial distance from the sun, the speed eventually becomes supersonic, i.e. larger than the local speed of sound in the corona. Near the earth's orbit, the solar wind blows at the incredible speed of about 400 kilometres per second. The distance from New York to Chicago can be covered in three seconds at that speed. I should emphasize that it is the high temperature of the corona giving rise to high pressure that is ultimately driving the solar wind. If we did not have a hot corona, then there would not be a supersonic solar wind.

Although the solar wind takes away very little mass from the sun, it is estimated that it takes away a considerable amount of angular momentum. The sun now takes about 27 days to go around its rotation axis. When the sun was young, it rotated much faster and its rotation period was considerably shorter. To understand how the solar wind removes angular momentum from the sun, we first need to figure out what the solar wind does to the sun's magnetic field. Although the strongest magnetic fields at the solar surface occur inside sunspots, we have also discussed magnetic fields outside sunspots (Section 7.2). At and below the solar surface, the magnetic field tends to be concentrated in magnetic flux tubes due to interactions with convection. Above the solar surface, however, magnetic field lines spread out and tend to fill up all space. The gas in the corona is filled with magnetic fields. The shape of the corona seen during the solar eclipse gives an indication of the structure of magnetic fields in the corona. A look at Figure 8.2 almost makes you feel that you are seeing the magnetic field lines of the sun. The appearance of the corona changes quite a bit from the sunspot maximum to the sunspot minimum. According to Alfvén's theorem of flux freezing, any moving plasma tries to carry the magnetic field with it. As a result of the solar wind trying to take the magnetic field with it, the magnetic field lines get stretched out. Figure 8.3 shows how the magnetic field lines would appear if we could view them from above the sun's pole. Suppose the sun is rotating in the anti-clockwise direction. Parts of magnetic field lines would go inside the sun and would be dragged with the sun's rotation, because of flux freezing. This would result in spiral shapes of magnetic field lines. These spiral magnetic field lines are known as *Parker spirals*—in honour of Parker who analysed this

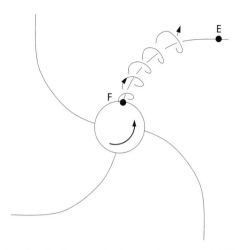

Figure 8.3 A sketch of Parker spirals, the spiral magnetic field lines of the sun stretched out by the solar wind. The thick curved arrow inside the sun indicates the direction of the sun's rotation. We also indicate how charged particles from the flare site *F* can reach the earth *E* if *F* and *E* are connected by a Parker spiral.

spiral structure of magnetic field lines in his famous 1958 paper proposing the solar wind. I have already pointed out that magnetic fields can exert some forces on the plasma (see the end of Section 4.6). A straightforward mathematical analysis shows that, as the solar wind plasma flows out, the spiral magnetic field lines exert forces on this plasma to make it rotate with the sun. As a result, the magnetic field lines transfer some of the sun's angular momentum to the outflowing plasma which carries it away. That is why we believe that the sun is losing angular momentum much faster than it is losing mass.

Now we are ready to address the question of how the charged particles accelerated to high speeds in solar flares reach the earth to produce polar aurorae. If these charged particles spread equally in all directions, then too few of them would come near the earth. But I have already pointed out that charged particles move in helical paths around magnetic field lines. Suppose a solar flare occurs at point *F* in Figure 8.3. The charged particles accelerated in this flare will move in helical paths around the Parker spiral starting from point *F*. If the earth *E* happens to lie near this Parker spiral as shown in Figure 8.3, then many of these charged particles will reach the earth to be eventually channelized by the earth's magnetic field to the geomagnetic poles, producing an aurora. On the other hand, if the earth were to lie far away from the Parker spiral emanating from the flare site, then the flare would not have much effect on the earth. In other words, not all solar flares affect the earth. The so-called *geo-effectiveness* of a flare depends on whether the earth is close to the Parker spiral starting from the flare site.

8.3 Bones of the Corona and the Secret of Heating

Suppose you heat a piece of iron to a very high temperature. At first, the iron becomes red-hot. On increasing the temperature further, it looks yellowish and eventually bluish-white. We know that yellow light has a shorter wavelength than red light and blue light has a still shorter wavelength. Basically we find that the piece of iron emits more and more radiation in shorter wavelengths as its temperature is raised. A mathematical statement of this is known by the rather curious name of *Wien's displacement law*—after Wilhelm Wien who won the Nobel Prize in 1911 for his studies of heat radiation. For those readers who are finicky, I should mention that, strictly speaking, this law applies only to a special kind of radiation known as blackbody radiation. But this law gives an approximate idea of the wavelength of light which will be emitted by a hot body at a certain temperature. When something is heated to a million degrees, it becomes a gas. An application of Wien's displacement law shows that the radiation coming out of this million-degree gas will have wavelengths close to a millionth of a millimetre. Only towards the end of the nineteenth century, were physicists able to produce radiation of such short wavelengths. Such radiation is called X-ray. If the gas in the corona is really as hot as a million degrees, we would expect X-rays to come out of the corona.

The earth's atmosphere happens to be opaque to X-rays. Even if an astronomical object emits X-rays, they will not reach us. Only if an X-ray detecting equipment is taken above the earth's atmosphere, can one expect to receive X-rays from astronomical sources. Nowadays it has become customary to put X-ray telescopes in satellites orbiting the earth, and many discoveries made with such X-ray telescopes have revolutionized astronomy—including providing the first convincing proof of the existence of black holes. It is remarkable that the Space

Age dawned within a few years of the realization that the corona is so hot. X-ray detecting instruments carried in rockets in the 1950s and 1960s indicated that X-rays were indeed coming from the solar corona. However, the real breakthrough came when NASA decided to put the Apollo Telescope Mount, a full solar observatory with various instruments, in the *Skylab*, the first US space station launched on 14 May 1973. Three teams of astronauts worked in the Skylab during 1973–74 carrying out observations of the sun. Solar astronomers saw X-ray photographs of the sun which seemed stunning at that time, but now may appear crude compared to the much sharper X-ray photographs taken from later space missions. Figure 8.4 shows an image of the

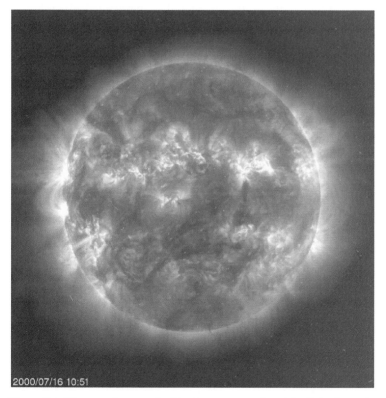

2000/07/16 10:51

Figure 8.4 The sun photographed in the extreme ultraviolet light from the space mission SOHO. This image was taken on 16 July 2000, around the time of a sunspot maximum. Credit: SOHO (ESA and NASA).

sun taken from the space mission Solar and Heliospheric Observatory (abbreviated to SOHO). This image was taken in 2000, at the time of the sunspot maximum, using radiation of somewhat longer wavelength compared to usual X-rays (called extreme ultraviolet). The surface of the sun appears dark, because a surface at a temperature of 6000 degrees does not emit X-rays or extreme ultraviolet. One can see loops in the corona which are sparkling like jewels. These loops, which are the hottest regions of the corona, are nothing but magnetic flux tubes connecting sunspot pairs, as sketched in Figure 2.8(a) or Figure 5.3(b). This becomes clear if one compares a figure like Figure 8.4 with an ordinary photograph of the sun taken at the same time. A loop is usually seen as a structure overlying two sunspots. A close-up of a coronal loop was shown in Figure 2.9, with its horseshoe-like structure clearly visible.

Whatever be the mechanism for producing heat in the corona, Skylab observations made it clear that this mechanism is most effective inside the magnetic loops, which are clearly hotter than the surrounding corona. The magnetic fields in the corona therefore must have something to do with the high temperature of the corona. One standard result of plasma physics is that, when a magnetic field is present in a plasma of low density, heat flows along magnetic field lines more easily than in the perpendicular direction. If heat is produced in some regions of a coronal loop, this heat will flow quickly along the loop making the whole loop light up; but the heat will not flow so easily to the gas outside the loop because that would require a heat flow perpendicular to the magnetic field. This is why the loops become much hotter structures than the surrounding gas and can be photographed in X-rays which they emit. The loops appear like the bones of the corona. We all know that X-rays are used to study the bones of our bodies. X-rays can pass through the flesh of our bodies unhindered and the bones, which are opaque to X-rays, cast shadows. X-rays are now also used to study the bones of the corona—the magnetic loops—though in an opposite way. Here these bones of the corona, instead of casting shadows in X-rays, emit X-rays themselves and can thus be photographed in X-rays.

Now at last we are ready to address the question of how the heat is produced in the corona to cause its very high temperature. The existence of magnetic fields in the corona means that there are currents in the corona which are giving rise to these magnetic fields. Now, we know that a current flowing through a wire makes the wire heated, since the current has to overcome the resistance of the wire and some energy is dissipated in this process causing the heat. Is it possible that the currents in the corona responsible for the magnetic fields are producing the heat while overcoming the electrical resistance in the corona? While the heat requirement of the corona is very low because of its extremely low density, the electrical resistance of the corona is also very low. If the currents flow through wide regions of the corona, simple estimates show that the heat produced is too little to explain the high temperature of the corona. Is there any way to enhance the heat production? One of the things we learn in high school physics is that, if the same current is made to flow through a thin wire and a thick wire, then more heat is produced in the thin wire. If the currents flowing through the corona can be made to flow through narrow regions of the corona rather than through wide regions, then much more heat would be produced.

In 1972 Eugene Parker wrote a brilliant and provocative paper arguing that the currents in the coronal loops are indeed forced to flow through fairly narrow regions.[9] Let me try to explain the main thrust of his argument. Figure 8.5(a) sketches what magnetic field lines may look like inside a coronal loop. What is shown in Figure 8.5(a) is like the simplest possible configuration of magnetic field lines. Parker realized that in reality the magnetic field configuration should be much more complicated. The magnetic field lines of the loop certainly continue below the solar surface, since magnetic field lines cannot end abruptly and presumably the configuration of the magnetic field underneath the

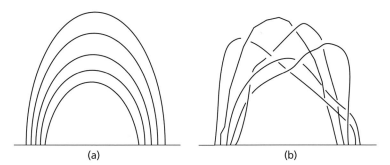

(a)   (b)

Figure 8.5 Sketches of magnetic field lines in a magnetic loop above the solar surface. (a) The simplest configuration one may naively expect. (b) The configuration resulting from random motions of footpoints.

surface looks as sketched in Figure 2.8(a). The point of the solar surface through which a magnetic field line goes beneath the surface can
be called the footpoint of the magnetic field line. Now, the solar surface
is the top of the sun's convection zone and the footpoints are always
disturbed by the random motions of convection. When the footpoints
move due the motions of convection, Alfvén's theorem of flux freezing
suggests that the footpoints will carry their magnetic fields with them.
Parker realized that, as a result of these random motions of footpoints,
the magnetic fields in the loop would get tangled up and would appear
as shown in Figure 8.5(b) rather than as in Figure 8.5(a). For such tangled magnetic fields, Parker proved a mathematical theorem that the
currents would be forced to flow through very narrow regions called
current sheets. In the scenario proposed by Parker, a coronal loop must be
full of such current sheets within which heat is produced to cause the
high temperature of the loop. This theory of Parker's remained highly
controversial for several years. When I was a PhD student, sometimes
special sessions would be organized in scientific conferences to debate
whether this theory was correct or not. People pointed out that Parker's
proof of the mathematical theorem he proposed was not very rigorous.
However, several numerical simulations performed in the following
years confirmed that the basic scenario proposed by Parker is sound,
although his proof may not be fully satisfactory. As footpoints are shuf-
fled around by random motions of convection, the overlying magnetic
fields get tangled up, leading to the formation of many current sheets
in the coronal loop within which heat is produced.

Here I should mention that another competing theory of coronal
heating began to be developed even before Parker formulated his theory of heat production through many current sheets. I have already
discussed in Section 4.3 that a magnetic field line has tension associated with it like a stretched string. When a stretched string is plucked,
we know that this tension causes waves to propagate along the string.
This is the basic principle underlying all stringed musical instruments.
Now let us suppose that a bunch of magnetic field lines get disturbed
as shown in Figure 8.6—very much like a plucked string. Such a disturbance in the magnetic field lines can be caused by a plasma motion
as indicated by the thick arrow. Exactly like what happens in a plucked
string, we expect the magnetic tension to give rise to waves along the
magnetic field lines. Hannes Alfvén, who derived the very important
theorem of flux freezing, carried out an analysis of the MHD equations

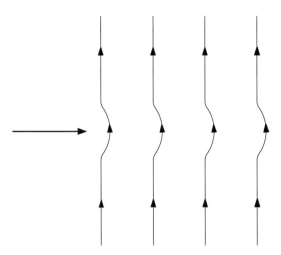

Figure 8.6 A sketch illustrating how an Alfvén wave is generated. Magnetic field lines are disturbed like a plucked string by a plasma movement indicated by the thick arrow.

in 1942 to show the existence of such waves propagating along magnetic field lines.[10] A wave of this kind is now called the *Alfvén wave*. All the magnetic field lines in the corona have parts going underneath the solar surface. The plasma motions due to convection taking place at and below the solar surface are expected to disturb the magnetic field lines, thereby launching Alfvén waves propagating upwards. Sometimes these Alfvén waves get mixed with ordinary acoustic waves to give rise to rather complicated waves. As the waves propagate through the corona, they get damped. During this damping process, the energy carried by the waves will be deposited in the corona in the form of heat. According to what is called the wave heating theory of the corona, this is how heat is produced in the corona to raise its temperature. Doing a full quantitative analysis of this process is extremely challenging. Bernard Roberts, Joseph Hollweg and Marcel Goossens have been prominent among the many scientists who studied theoretically how Alfvén waves propagate through complicated magnetic structures and get damped.

There was a time when the wave heating theory and Parker's theory of current sheets were regarded as two rival models of coronal heating. Now many solar physicists think that both the theories are applicable in different situations. Since the coronal loops are much hotter than the surrounding regions of the corona, it is quite plausible that the loops have a different heating mechanism. Present thinking is that the loops are heated by many current sheets formed within them as envisaged by Parker. On the other hand, the surrounding regions have open magnetic field lines stretched out by the solar wind. These regions are presumably heated by Alfvén waves propagating along the magnetic field lines. This is a broad-brush qualitative scenario of how the corona gets heated to very high temperatures. The challenge now is to do a detailed quantitative analysis of the amount of heat generated by the processes outlined above and to compare with observational data. As more and more high-quality data for the corona come in from various space missions, this has now become a highly active research field. In a few years, we shall hopefully be sure if our current ideas of coronal heating are really correct.

8.4 When Magnetic Field Lines Swap Partners

According to Parker's theory, many current sheets form inside a coronal loop due to motions of its footpoints. The study of current sheets has become a very important area of research in plasma physics, since there are many situations when current sheets occur inside plasmas. Let us discuss a little bit more what current sheets really are—especially because this discussion will provide a clue to answering one of the basic questions we are interested in: why solar flares occur.

Figure 8.7 shows a region OP on two sides of which the magnetic field lines BC and $B'C'$ have opposite directions. From the basic equation of electromagnetism connecting magnetic fields and electric currents (called Ampere's equation), it follows that OP should be a region of strong current perpendicular to the page of this book—a current sheet. This strong current working against the electrical resistance would produce heat there. Now, we know from Alfvén's theorem that the magnetic field is frozen in the plasma when the effect of resistance is negligible. We expect the effect of resistance to be important only around OP where there is a strong current. The magnetic field will be frozen elsewhere, but not around OP. As the strong current through OP would tend to decay due to the effect of resistance, the parts BC and $B'C'$ of magnetic field lines would decay away. But we know that a magnetic field line cannot

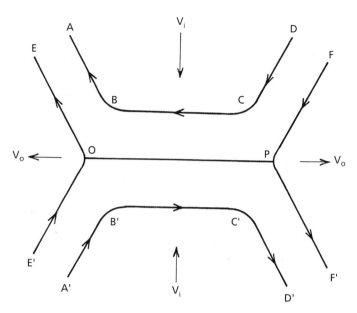

Figure 8.7 Sketch of a current sheet within which magnetic reconnection takes place.

end abruptly somewhere in space. If the part BC of the magnetic field line $ABCD$ decays away, what will happen to the rest of this field line? We certainly cannot expect the field line AB to end at B.

We know that magnetic fields have pressure associated with them. When the magnetic fields around OP decay, the pressure will decrease there due to the disappearance of the magnetic field. As a result, the plasma above and below OP will be sucked towards OP at speed V_i indicated in Figure 8.7. Since this incoming plasma will carry the magnetic field with it, the magnetic field lines $ABCD$ and $A'B'C'D'$ will be brought near to each other. When the parts BC and $B'C'$ decay away, it will be natural for the parts AB and $A'B'$ to get joined together to make a continuous magnetic field line looking somewhat like the field line EOE'. In exactly the same way, CD and $C'D'$ will get connected in a continuous field line. We find that parts of the magnetic field lines are swapping partners. Initially AB and CD were parts of the same magnetic field line: they were partners. After this field line is sucked to the current sheet region OP, AB will take on $A'B'$ as its partner and CD will take on

$C'D'$ as its partner. This process of parts of magnetic field lines swapping partners is known as *magnetic reconnection*. The reconnected magnetic field lines EOE' and FPF' have tensions which will try to straighten out these field lines. In this process, these field lines will get pushed away from the central region with speeds V_0 indicated in the figure.

Peter Sweet, Eugene Parker and Harry Petschek figured out the physics of current sheets in the late 1950s and early 1960s. It is quite a complex process and not easy to analyse mathematically. I hope that my description of this process gives you an idea of what is involved. Magnetic fields from both sides of the current sheet are sucked into the current sheet. There the magnetic field lines swap partners and the reconnected field lines are pushed away from the central region. For the parts of magnetic field lines which decay away due to resistance, the energy of the magnetic field gets converted into heat. We have pointed out that many current sheets form within coronal loops and produce heat there. But is it possible for much larger current sheets to occur in the corona?

We believe that solar flares are caused by current sheets of gigantic size. In fact, the idea of current sheets was first proposed in the late 1950s to provide an explanation for solar flares. We know that magnetic flux keeps coming out to the corona from underneath the solar surface due to magnetic buoyancy. Suppose some new magnetic flux comes out in a region which had some old magnetic flux. If the old and the new magnetic fluxes are oriented in different directions, then we may expect a gigantic current sheet between them. That is one of the possible mechanisms for producing a flare. In the current sheet responsible for the solar flare, a large amount of magnetic energy has to be converted quickly into heat, light and other forms (like the energy of fast moving charged particles which get accelerated in solar flares). Solar flares naturally occur in regions where there is a large store of magnetic energy. The strong magnetic fields above sunspots are the favoured regions for the occurrence of solar flares. Since a large flare involves a huge amount of energy released within a very short span of a few minutes, the challenging question before the theoretical solar physicist is to figure out how the magnetic energy is converted so fast in the current sheet. First estimates suggested a much slower rate of magnetic energy conversion. However, quite a lot of research has been done on this subject in the last few years and it has at last been possible to construct detailed models of solar flares. The groups of Peter Sturrock, Eric Priest and Kazunari Shibata have made tremendously important contributions to this field.

Since this is a highly technical subject, we cannot go into more details of it here. The basic scenario we have is that a flare involves a rapid conversion of magnetic energy to other forms in a huge current sheet formed in the corona above sunspots.

8.5 Arrivals at the Station Earth

I have repeatedly made use of Alfvén's theorem of flux freezing in our discussions. One of the most important corollaries of this theorem is the following. A plasma blob initially outside a magnetic field cannot easily enter the region of this magnetic field. Suppose the plasma blob initially had no magnetic flux. If it could get inside the magnetic field, then it would thereafter acquire some magnetic flux passing through it. This would contradict Alfvén's theorem that the magnetic flux through the blob of plasma should not change.

The region around the earth filled with the earth's magnetic field is called the earth's *magnetosphere*. The solar wind plasma cannot easily get inside the magnetosphere for the reason given in the previous paragraph. So the front side of the magnetosphere towards the sun acts like a shield diverting the solar wind to flow around the magnetosphere, as shown in Figure 8.8. If the space around the earth were empty, as believed until the discovery of the solar wind, then the magnetosphere of the earth would have been completely symmetric around the earth's magnetic axis. But the solar wind squeezes the magnetosphere on the day side and makes it extended on the night side. Once we have this basic idea about the structure of the magnetosphere, we are at last ready to discuss how geomagnetic storms may arise.

In the 1930s, about a quarter of a century before Parker predicted the solar wind, Sydney Chapman and Vincenzo Ferraro proposed a rather remarkable idea how solar flares may give rise to geomagnetic storms.[11] They suggested that a violent explosion like a solar flare may throw away a large chunk of plasma from the sun. This chunk of plasma would then move through what was at that time considered to be empty space. When this chunk of plasma reaches the earth, it would squeeze the earth's magnetosphere, changing the positions of the earth's magnetic field lines and causing a sudden variation in the earth's magnetic field. This was a really prophetic theoretical conjecture. It was only many decades later that observational data started to

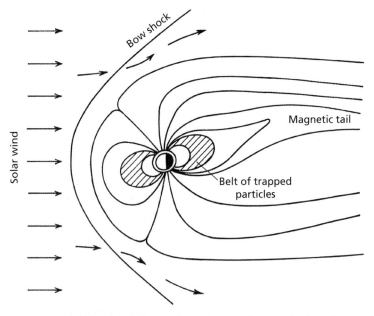

Figure 8.8 The sketch of the earth's magnetosphere with the solar wind flowing around it.

come in support of the idea that a solar flare may indeed throw away a large chunk of plasma.

If a large chunk of plasma is suddenly thrown up from the sun, we may expect to see it as a bright structure in the corona. Under normal circumstances, the glare of the sun's disc will not allow us to see such bright structures in the corona. Only through a coronagraph, may we expect to see these bright structures which eventually get carried away with the solar wind. Such a chunk of plasma thrown up from the sun is called a *Coronal Mass Ejection*, abbreviated to CME. Figure 8.9 shows a CME observed from space by the coronagraph on board the Solar and Heliospheric Observatory (SOHO). In the early 1970s Richard Tousey's team discovered CMEs with the help of a rather primitive coronagraph flown in space. Ever since SOHO was launched in 1995, the coronagraph on board has been regularly finding CMEs, which has allowed solar astronomers to build up a large catalogue of CMEs.

Although we often see a CME following a solar flare, there is not a 100% one-to-one correspondence between flares and CMEs. Sometimes

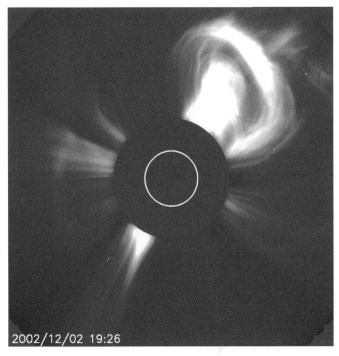

2002/12/02 19:26

Figure 8.9 A coronal mass ejection (CME) seen on 2 December 2002 through the coronagraph LASCO C2 on board the space mission SOHO. Credit: SOHO (ESA and NASA).

there are flares which do not produce CMEs and sometimes there are CMEs which do not seem to follow flares. In a popular science book like this, we shall not get into all these intricacies. We shall take the point of view that a CME is a chunk of plasma with magnetic fields inside it, which is thrown up from the sun by some kind of explosion—a flare being the most likely cause. Now, when magnetic fields come out of the solar surface, they usually have parts of field lines going below the surface. Such magnetic connectivities may make it difficult for a magnetic structure to break away from the sun. However, we have seen that magnetic reconnections may take place in a current sheet like what occurs in a flare, allowing magnetic field lines to swap partners. As a result of magnetic reconnections, a magnetic structure may become detached from the sun and may break away carrying the plasma in which it is embedded. A CME is presumably a chunk of plasma with such a magnetic structure embedded in it. Scientists like Boon Chye Low and Spiro Antiochos have studied the physics of how magnetic structures may break away from the sun.

Since magnetic fields are responsible for causing solar flares and CMEs, it is no wonder that flares and CMEs occur more frequently during sunspot maxima when magnetic fields are more abundant on the solar surface. The statistics of flares and CMEs show a clear 11-year cycle in tandem with the sunspot cycle. When CMEs first start, different CMEs move with different speeds—some moving slowly and some moving more rapidly. Eventually they all move with the solar wind at its speed. As I already pointed out, a CME is a chunk of plasma with a magnetic structure embedded in it. According to Alfvén's theorem of flux freezing, the magnetic structure gets carried with the chunk of plasma—expanding and becoming larger in size as the CME moves farther to regions of lower density. A geomagnetic storm occurs when the CME hits the top of the earth's magnetosphere, in essentially the same way as Chapman and Ferraro envisaged. As the CME impinges on the magnetosphere like a battering ram with its huge kinetic energy, the magnetosphere is pushed downwards—moving the magnetic field lines and thereby changing the earth's magnetic field. Since the regions around the geomagnetic poles have the strongest concentration of earth's magnetic field lines and the relative openness of the field lines there allows solar disturbances to reach closer to the earth, it is no wonder that geomagnetic storms have their maximum strength around the geomagnetic poles. Under normal circumstances, the earth's magnetosphere extends to a distance of about 70,000 kilometres on the day side. Under the impact of a large CME, the top of the magnetosphere may get lowered quite a bit. Lord Kelvin was against the idea of a solar cause for geomagnetic storms because he thought that the space between the sun and the earth was empty and any magnetic disturbance would spread from the sun in the way that a magnet's effect usually spreads. If that were the case, then even the most violent magnetic explosion on the sun would create barely a stir in the earth's magnetic field. Lord Kelvin had no idea that many years later it would be found that there is a continuous plasma flow from the sun, encompassing the earth's magnetosphere, and that magnetic structures from the sun can be carried along this flow without a reduction in magnetic flux according to Alfvén's theorem. I mentioned that Lord Kelvin used his address as President of the Royal Society to attack those who were looking for solar causes behind geomagnetic storms (Section 1.1). It is curious to note that Edward Sabine, the early proponent of the solar origin of geomagnetic storms, had occupied this chair (President of the Royal Society) only a few years before Lord Kelvin.

If a solar flare as strong as the 1859 Carrington flare were to occur today, it may render electrical power grids around much of the world dysfunctional. The total cost to the world economy may be comparable to the gross domestic products of some countries. Since we cannot prevent such a calamity, the best we can try to do is to predict its occurrence at least a few hours in advance so that suitable precautions can be taken. Not all solar flares affect the earth. So, if we were to start taking expensive precautions after each and every solar flare, that would be a rather inconvenient way of doing things. We need to figure out which solar flares are likely to cause maximum havoc on the earth. I have already mentioned that, if the flare site and the earth are connected by a Parker spiral, then accelerated charged particles can reach the earth to produce an aurora. A parcel of gas in the solar wind would also flow along the Parker spiral. So CMEs are also expected to follow Parker spirals—after they start moving with the solar wind. If a flare site is connected with the earth through a Parker spiral, then a CME produced by that flare is likely to hit the earth. The need of the hour is continuous monitoring from space to check whether a big CME is proceeding towards the earth. One NASA mission precisely designed for this purpose is STEREO, launched in 2006. It consists of two identical solar observatories—one ahead of the earth in its orbit around the sun and the other behind the earth. By combining observations from two different points in space, STEREO is able to reconstruct three-dimensional views of CMEs, showing us how they move.

8.6 A Grand Landscape

If the sun did not have magnetic fields, it would have been just a sphere of hot gas without any prominent features breaking its spherical symmetry. Sunspots, flares, the hot corona and the solar wind are all caused by magnetic fields. All these things are often collectively called solar activity. If the sun did not have these activities, it would certainly be a much less interesting object.

A branch of science in which different important topics are interconnected through some unifying principles always possesses a special kind of intellectual appeal. This is certainly true for the study of solar activity. The magnetic field is the unifying thread and all the various phenomena connected with solar activity can be explained by applying the principles of MHD. What is more, one phenomenon leads to another and then that another phenomenon leads to the next through a chain of causes and effects. We can begin with the dynamo process which produces the magnetic field in the interior of the sun. Then this magnetic field floats to the surface by magnetic buoyancy and gives rise to sunspots. The magnetic fields above sunspots can have current sheets within them producing the hot corona and the occasional flare. It is the hot corona which drives the solar wind. The study of all these interconnected phenomena through the principles of MHD is called solar MHD. At a time when the various interconnections in this field were just becoming clear, the publication of an influential textbook, Eric Priest's *Solar Magnetohydrodynamics*, in 1982 did a lot towards consolidation of the field. I was a PhD student at the time of its publication. I still remember how excited I was when I got the book in my hands for the first time. For days I pored over its pages to develop a broad perspective of our field. Generations of students in our field have learnt their basics from this textbook.

I got initiated into this field when the various interconnections in this field had just been established. I had not been a first-hand witness of the period when the interconnections were being discovered. But I get a pretty good feeling of how the field emerged when I look at the volume *The Sun*, published in 1953 under the editorship of Gerard Kuiper. Several eminent scientists of that time contributed chapters on different aspects of solar physics in that volume, which was justly regarded as a definitive scholarly volume surveying the whole of solar physics at it was then. I have already mentioned that up to the middle of the twentieth century the study of solar activity was a Baconian enterprise in which scientists collected and organized data without any underlying concepts with which to weave them together into a theoretical framework (Section 1.3). The volume edited by Kuiper shows the field still in that state. Several chapters in the volume discuss various kinds of data connected with solar activity available at that time, with very little attempt at any theoretical interpretation. MHD was already emerging as a new scientific discipline and Thomas Cowling of Cowling's anti-dynamo theorem fame (who happened to be the PhD supervisor of Eric Priest, the author of *Solar Magnetohydrodynamics*) contributed a chapter on MHD to that volume. Although there was a hope at that time that

MHD might eventually explain many aspects of solar activity, it was still a hope rather than a demonstrated fact. Cowling mainly reviewed the basic principles of MHD, which were not widely known to solar researchers at that time. Towards the end of the chapter, he briefly mentioned some of the theoretical ideas of that period on sunspots and their cycle. Most of these ideas, discarded long ago, now appear strange and fanciful to us. Hopefully our present-day ideas on these topics will stand the test of time and will not appear so strange to coming generations!

If we compare this 1953 volume *The Sun* with Eric Priest's *Solar Magnetohydrodynamics* published in 1982, we can realize the enormity of the progress achieved in this field over the intervening three decades. While Priest's book has become somewhat outdated by now due to rapid developments in this field, the basic theoretical framework presented in that book still stands in full glory. We can assess the evolution of this field in the light of certain ideas developed by the philosopher of science Thomas Kuhn in his famous book *The Structure of Scientific Revolutions*. According to Kuhn, a scientific field may occasionally have scientific revolutions when completely new paradigms for research emerge. After a scientific revolution, scientific progress continues by means of what Kuhn calls normal science, during which the paradigms established during the scientific revolution are used successfully to develop a much deeper understanding of the field. I would like to suggest that the intervening period between the volume *The Sun* and Eric Priest's book has been a period of scientific revolution in our understanding of solar activity. New paradigms of research emerged during this period and showed how the principles of MHD can reveal a grand landscape in which different aspects of solar activity appear interconnected. After this scientific revolution, however, an equally exciting period of normal science has followed and is still continuing, with the basic paradigms being applied to give us a fuller understanding of solar activity. In a non-technical book like this, it is more difficult to convey the excitement of normal science, which is usually the normal state of research in a scientific field and is no less exciting to the practitioners in the field. In the case of at least one topic—solar dynamo models—I have given an account of how Parker's 1955 paper provided only the basic framework and how some of us then developed more detailed models guided by this basic framework.

One of the amazing things in the development of solar MHD is that many of the paradigms of this field came from just one man: Eugene Parker. It was he who first showed how the dynamo mechanism generates magnetic fields. Then he was the person to show how magnetic fields from the solar interior may rise due to magnetic buoyancy to create sunspots. Again it was Parker who figured out how magnetic fields above sunspots may develop current sheets within them and give rise to the hot corona. Perhaps his most dramatic contribution was to show how the hot corona would drive the solar wind. He strode over the field of solar MHD like a colossus. There are not many such examples of one person developing virtually all the key concepts in a scientific field. Parker started publishing his first papers at about the time when the volume *The Sun* came out—a time when MHD had just emerged as an important branch of plasma physics and when the first efforts of applying MHD to solar phenomena had begun. In the next three decades, Parker's work established the broad theoretical framework of the field. In an autobiographical piece written in 1985, Thomas Cowling frankly and generously wrote: 'I have been unable to vie with Parker in his subsequent meteoric career.'[12]

I have often wondered, if one man named Eugene Parker had not existed, then how would our field have evolved. This is the type of question to which one can never get an empirical answer. So I can only guess. Probably many others collectively would have made all the discoveries which Parker alone made. But the field definitely would have taken many more years to mature. On several occasions, Parker was years ahead of his time. His theory of the turbulent dynamo, his prediction of the solar wind and his idea of current sheet formations in the corona took years to be understood and appreciated. Had Parker not existed, I am reasonably certain that the solar wind would not have been predicted theoretically, but would have been first discovered from space missions. Then probably some day some theorist would have figured out the cause behind it. It is possible that some of the inter-relations amongst different aspects of solar activity which were established by the time of Priest's book would still elude solar physicists and a much more protracted 'scientific revolution' might have continued till today. Whether that would have been more fun or less fun certainly depends on your point of view.

9

We Look Before and After

9.1 What the Trees and the Ice Tell Us

No two sunspot cycles are exactly alike. Each cycle has its own distinctive personality and usually holds some surprises for us. Take another look at Figure 1.5. While several cycles were missing in the period 1640–1720, a few successive cycles around the middle of the twentieth century were stronger than usual. Just as a prolonged period of missing cycles is called a 'grand minimum', a period of several successive strong cycles is called a 'grand maximum'. Apart from the occurrences of grand minima and grand maxima, individual cycles keep varying in strength. The durations of cycles also vary. While the average duration of sunspot cycles is about 11 years, there have been cycles as short as 8 years and as long as 14 years.

We have discussed in Chapter 7 how the flux transport dynamo model explains the sunspot cycle. Now we come to the next question. Why are sunspot cycles not all exactly alike? What causes the irregularities of the cycles? Coupled with this question, there is another extremely important issue. If we can figure out what causes irregularities in sunspot cycles, will that give us the power to predict future cycles? Shall we be able to say whether the next cycle is going to be strong or weak? Long or short? This is not simply a theoretical question arising out of our intellectual curiosity, but has tremendous practical significance as well. A stronger cycle means that there will be a larger number of solar disturbances during the cycle which are likely to affect us earthlings. A stronger cycle shortens the working life of a man-made satellite and makes air routes over magnetic poles dangerous more frequently. Somebody who is a senior manager in an airlines company or is in the business of launching communications satellites would surely like to know whether the next sunspot cycle will be strong or weak. Can we predict this reliably?

If the irregularities of the sunspot cycle are really 'irregular', then it will be very hard to make any meaningful predictions. However, a careful analysis of the irregularities of the sunspot cycle enables us to discern some patterns within these irregularities. It is these patterns which provide us with clues about how the irregularities arise, giving us the hope that some limited prediction may be possible. To proceed along these lines, we would like to have as much statistical data on the irregularities of sunspot cycles as possible. However, we have sunspot records left by astronomers only for about four centuries and these records also become less reliable as we go back earlier than the nineteenth century. Is there any way to reconstruct the history of sunspot activity before Galileo turned his telescope to look at sunspots? Only a few years ago, scientists managed to come up with some really ingenious methods for reconstructing the history of sunspot activity at earlier times.

Probably all readers of this book will have heard of radiocarbon dating, although many may not know the scientific principle behind it. We have already discussed cosmic rays (Section 8.1), which are fast-moving charged particles bombarding the earth all the time. When some of these particles collide with atomic nuclei in the air, one of the end products is an isotope of carbon C-14, which is radioactive with a half-life of about 5700 years. This means that, if some nuclei of C-14 are kept together, half of them will decay (into nitrogen nuclei) in 5700 years. Because of the continuous production of C-14 in the atmosphere, it is usually assumed that a certain fraction of carbon atoms in the atmosphere (usually present within carbon dioxide molecules) would have C-14 in their nuclei. These C-14 nuclei would get absorbed into plant tissues during photosynthesis. Once photosynthesis stops, the C-14 nuclei will simply undergo radioactive decay. By analysing the C-14 content in a block of wood or in a piece of cloth, one can figure out its age.

In carrying out radioactive dating, it was at first assumed that the fraction of C-14 in the atmosphere is in a steady state and does not change with time. However, about a decade after Willard Libby developed the radiocarbon dating technique (in the mid-1940s), careful measurements by Hessel de Vries suggested that the C-14 concentration in the atmosphere in the late seventeenth century was slightly higher. Such fluctuations in C-14 concentration are now called *de Vries fluctuations* and have explained many puzzles in radiocarbon dating. For example, on assuming a constancy of C-14 concentration in the atmosphere, the mummy of an Egyptian Pharaoh was dated to be older than his grandfather's mummy! Only after taking account of de Vries fluctuations, could such anomalies be resolved. Many think that de Vries might have shared the 1960 Chemistry Nobel Prize with Libby if he had not committed suicide in 1959 after murdering the woman whom he loved but who wanted to marry another man (a variation on the theme of the Richard Carrington tragedy?).

John Eddy, whose famous 1976 paper on the Maunder minimum we have already mentioned (Section 2.2), noted in that paper that the enhancement of C-14 in the atmosphere discovered by de Vries occurred at the same time as the Maunder minimum. Explanation of this is not difficult. Cosmic ray particles are deflected by the magnetic field of the solar wind and have to work their way through this magnetic field to reach the earth. Eugene Parker and Randy Jokipii in the 1960s developed the theory of how cosmic ray particles make their way through the labyrinthine magnetic fields in the solar wind. When there are no sunspots, presumably the general magnetic field of the sun is weak and this must be true for the magnetic field of the solar wind as well. In such a situation, cosmic rays would reach the earth more easily and produce more C-14 nuclei. In fact, in the 1950s Scott Forbush noticed that the intensity of cosmic rays was slightly reduced during the maximum of the sunspot cycle. Even after a major solar flare, the cosmic ray intensity drops temporarily. Such decreases in the cosmic ray intensity are known as *Forbush decreases*. We basically conclude that there would be less C-14 in the atmosphere when there are more sunspots and there would be more C-14 in the atmosphere when there are fewer sunspots. In other words, if we can determine the atmospheric concentration of C-14 at various times in the past, we can reconstruct the history of sunspot activity.

How can this be done? Sadly, the best method is to cut down a very old tree. One ring forms in the trunk of the tree every year. If one can date the tree rings, then an analysis of the contents of these tree rings provides information about C-14 concentration in the past years when these tree rings formed. It seems that the late medieval age (twelfth and thirteenth centuries) was a time when the C-14 concentration was slightly lower, indicating that this must have been a time when sunspot activity was particularly strong—an opposite of the Maunder minimum.

Cosmic rays produce other radioactive isotopes as well. One such isotope with a much longer half-life than C-14 is an isotope of beryllium Be-10. Its longer half-life enables scientists to carry out radioactive dating over a much longer time. Just as in the case of C-14, we expect an increase or a decrease in the atmospheric concentration of Be-10 to correspond respectively to a decrease or an increase in sunspot activity. How can we find out the past concentration of Be-10 in the atmosphere? In the ice fields of Greenland and Antarctica, an ice layer deposits every year. By drilling a hole in the ice field and extracting a long column of ice, trained scientists can figure out which layer of ice formed how long ago. By analysing an ice layer formed thousands of years ago, it is possible to determine the composition of the atmosphere (especially the Be-10 concentration) at that time. Such a polar ice core is virtually a time capsule, archiving the history of how our atmosphere has evolved over several millennia.

In recent years, Jürg Beer and Ilya Usoskin have been the leading scientists involved in reconstructing the past history of sunspot activity. By now it has been possible to extend this history over the entire geological period of the Holocene, which has lasted for the last 11,000 years. Especially, it has been found that a grand minimum of sunspot activity like the Maunder minimum is not something unique. It is estimated that there had been about 27 such grand minima over the entire Holocene.

9.2 Blow Hot and Blow Cold

I mentioned that the Maunder minimum was an epoch of unusually cold climate—the little ice age (Section 2.2). Presumably a fall in the strength of sunspot activity causes a cold climate on the earth. We expect the opposite also to hold. An increase in sunspot activity is likely to produce a warm climate. I pointed out in the previous section that the late medieval age (twelfth and thirteenth centuries) was a time of increased sunspot activity, as indicated by a decrease in the atmospheric C-14 concentration at that time. This suggests that the late medieval age was an epoch of warm climate. Now, the late medieval age happened to be a very fascinating time in history. Large human populations settled in many of the north European countries during this period; agriculture and trade flourished; many cities and towns were established. This was also the period when some of the great cathedrals of Europe were built. Maritime activity in the north Atlantic was also at its peak during this time. The Vikings sailed to Greenland. Presumably a mild and temperate climate stimulated all these human activities.

Historians know that a drought or a crop failure could lead to the downfall of a mighty empire. Rises and falls of many empires can be linked to changes in the climate pattern. Until human beings started contributing to global warming after the Industrial Revolution, sunspot activity appears to have been the main cause of climate changes at earlier times—the world being cooler during reduced sunspot activity and warmer during enhanced sunspot activity. Probably many major events in history have connections with sunspot activity. Research is just beginning in this fascinating field. A more complete reconstruction of sunspot activity in the last few millennia may enable historians to answer many questions which have remained unanswered so far.

We now come to the crucial question: why does the temperature at the surface of the earth increase and decrease with the increase and decrease in sunspot activity? This is so far a very ill-understood subject and some of the prevalent theoretical ideas are highly controversial. Still let me give a brief account of where we stand in this field.

Since sunspots are cool regions on the surface of the sun, you may naively expect that the total energy output of the sun would be less when there are many sunspots. This would suggest that the earth would cool when there are many sunspots—the opposite of what is found. One of the most surprising discoveries in solar astronomy in the last few decades is that the sun is actually brighter when there are many sunspots! One cannot measure the brightness of the sun very accurately from the surface of the earth, since the varying atmospheric conditions would badly affect the results of such measurements. However, from the late 1970s, radiometers carried on board several satellites have been continuously monitoring the energy flux from the sun. It appears that the energy flux of the sun increases by about 0.1% during the sunspot maximum. Figure 9.1, which presents the data of solar energy flux (called solar irradiance) variation with time along with the sunspot number, makes this very clear.

From the theoretical point of view, this is a very puzzling result. The dynamo process giving rise to the sunspot cycle is confined within the convection zone in the outer part of the sun, whereas the energy is generated by nuclear reactions in the central core of the sun. It is highly unlikely that the energy generation rate would change with the sunspot cycle. The magnetic fields in the convection zone can at most suppress or divert the heat flow through the convection zone. They have to do this in such a way that the sun becomes brighter when there

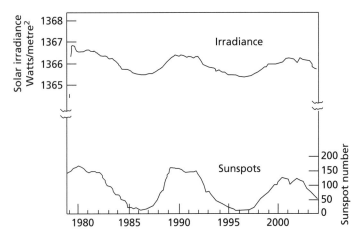

Figure 9.1 The upper curve is the variation of solar irradiance with time, whereas the lower curve depicts the sunspot number for comparison.

are many sunspots. So far, we do not have a fully convincing theoretical model to explain how this happens. There are some theoretical ideas which appear more like lame afterthoughts following the discovery that the sun becomes brighter during sunspot maxima. I do not intend to discuss these theoretical ideas here, because to me they do not seem to have the ring of truth.

The sun's brightness goes up and down in tandem with the sunspot number. Let us merely take this as an experimental result for which we do not have a proper theoretical explanation. We now come to the question of whether this variation of the sun's brightness can explain such climatic events as the little ice age during the Maunder minimum. Certainly the energy flux from the sun is the ultimate driver of the earth's climate. Over the past few decades, climate scientists have developed what are called global circulation models. In these models, one calculates the circulations which would be set up in the earth's atmosphere and oceans due to the energy flow from the sun and thereby finds out what the climate should be like. We would like to know the amount by which the earth's average temperature would fall when the energy output from the sun decreases by 0.1%. Since the global climate models are very complex and still have many uncertainties, it is difficult to find an unambiguous answer to this question. Presumably there is some response time for the earth's climate to change when the energy output from the sun changes. If this response time is of the order of a few years, then the variation in the sun's brightness with the sunspot cycle of 11 years may not produce much change in the earth's climate. Only when the sunspot activity remains exceptionally low or exceptionally high over several cycles, does the earth's climate system have time to respond.

Extrapolating the data shown in Figure 9.1, one may conclude that during the Maunder minimum, when the sunspot number was close to zero for much of the time, the sun's brightness would have dropped and that caused the cold climate. On the other hand, during the late medieval age (twelfth and thirteenth centuries), probably the sunspot number tended to be on the higher side and the sun was brighter, causing a warm spell on the earth. While many of the details may be missing, such arguments seem plausible. However, there is now a twist to the story due to some ideas propounded by Henrik Svensmark, a Danish scientist, from the late 1990s. I now come to these ideas.

When cosmic rays pass through the atmosphere, they ionize some air molecules. Svensmark and his colleagues argue that these ionized molecules act as centres around which water vapour condenses to form water droplets. In other words, when there are more cosmic rays, more clouds will form. Remember that the cosmic ray intensity goes up when there are fewer sunspots. So there must be more clouds when there are fewer sunspots. According to Svensmark, the clouds reflect a part of the sunlight back into space and thereby try to make the earth cooler. We thus have a second theoretical argument on how the earth may become cooler when there are fewer sunspots. There would be more cosmic rays producing more clouds, which reflect away more sunlight.

On the face of it, this certainly appears to be an attractive theoretical hypothesis. You might expect that the variation of the sun's brightness and the mechanism proposed by Svensmark would both affect the earth's climate. After all, both of them contribute in the same direction. When there are fewer sunspots, both of these mechanisms suggest a cooler climate. You might conclude that the little ice age during the Maunder minimum was caused by the combined effect of both. However, Svensmark has got into the eye of a storm by claiming that his mechanism alone provides a *complete* explanation of not only all the climatic variations before the Industrial Revolution, but also all the variations during the last 100 years. According to him, the variation of the sun's brightness has a minimal effect on the earth's climate. He also claims that man-made carbon dioxide has absolutely no effect on climate and there is no such thing as global warming caused by greenhouse gases. We can keep on producing as much carbon dioxide as we wish without causing any damage to our climate system. According to Svensmark's arguments, the higher world temperature in recent decades is entirely caused by enhanced sunspot activity. There have been fewer cosmic rays reaching the earth and fewer clouds to reflect sunlight away. It is true that sunspot activity was very high around the middle of the twentieth century. But that phase is over and the last two sunspot cycles have again been weaker. So, by Svensmark's argument, the global temperature should not rise any longer. Svensmark actually claims that this is the case: global warming is over and the earth has started cooling! Since I am not an expert on this subject, I can only say that such an assertion is completely at odds with what is found by the Intergovernmental Panel on Climate Change (IPCC).

Svensmark teamed up with the veteran science writer Nigel Calder to write a popular science account of his research: *The Chilling Stars*. You can read this book if you want to learn more about Svensmark's ideas. In spite of a glowing endorsement from Eugene Parker in the Foreword of the book, I personally find the book to be not well written and full of sloppy arguments. I give one example. While attacking Svensmark's critics, Svensmark and Calder write:

> A problem with climate science in general is that the system controlling events at the Earth's surface is elaborate enough for theorists to play endless games with it, moving ice, water, air and molecules like chessmen, to explain anything they like.[1]

Interestingly, Svensmark wants this remark to be applicable only to his critics. As far as he himself is concerned, in spite of this 'problem with climate science in general', Svensmark thinks that he has arrived at the final truth that the increase in global temperature in recent decades is entirely caused by sunspot activity and there is no such thing as global warming due to man-made carbon dioxide.

At present the majority of scientists with whom I have discussed this matter hold the view that the sun was the major driver for climatic change in historical times until the Industrial Revolution, but the global warming of the twentieth century has been caused by man-made greenhouse gases. While science does not exactly function like a democracy and there is no guarantee that the majority view will ultimately prove correct, this is where things stand now. Svensmark's discovery of the connection between cosmic rays and clouds is certainly an important discovery, even if we do not accept his extreme views. The claim that the theoretical warming based on Svensmark's mechanism precisely matches the actual warming in recent times appears to be spurious. Such precise theoretical calculations are hardly possible in climate science.

9.3 Searching for Correlations

Now we come back to the question of what causes the irregularities of the sunspot cycle. As I have already pointed out, one can get important clues by looking for patterns or 'regularities' within the irregularities. Take another look at Figure 2.11. You may notice that the polar field during the sunspot minimum around 1997 was somewhat weaker than the polar field during the previous sunspot minimum around 1986. This weaker polar field was followed by the sunspot cycle 23, which was weaker than the previous cycle 22. Do we expect a weak polar field at a sunspot minimum to be followed by a weak sunspot cycle and a strong polar field to be followed by a strong sunspot cycle? The best way of answering this question is by producing what scientists call correlation plots. One such plot is shown in Figure 9.2(a), in which the horizontal

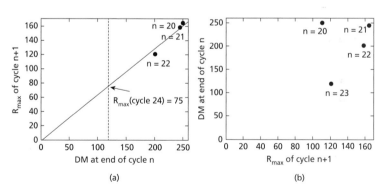

Figure 9.2 (a) A correlation plot between the polar field DM at the end of the cycle n and the strength R_{max} of the next sunspot cycle $n + 1$. (b) A correlation plot between the strength R_{max} of the sunspot cycle n and the polar field DM at the end of this cycle n.

axis corresponds to the strength of the polar field at the end of the cycle n ('DM' marked on the axis means dipole moment, which is a measure of the polar field) and the vertical axis corresponds to the strength R_{max} of the next sunspot cycle $n + 1$. From a plot like Figure 2.11, we can now read out the value of the polar field at the end of the cycle $n = 21$ and the peak value of the next sunspot cycle $n + 1 = 22$. These two values together are represented by the point marked 21 in Figure 9.2(a). Since we have reliable polar field measurements only from the mid-1970s, we have only three data points in Figure 9.2(a). However, all the three data points lie extremely close to a straight line. In scientific jargon, we say that there is a correlation between the polar field at a sunspot minimum and the strength of the next cycle. Since this conclusion is based on only three data points, its statistical significance is not too high and we cannot yet be very confident of this correlation. However, if this correlation really exists, then we have a powerful method for predicting future sunspot cycles. Suppose we assume that all future data points will also lie close to the straight line in Figure 9.2(a). Then, once we know the value of the polar field at a sunspot minimum, we can figure out the strength of the next cycle that would make the data point lie on the straight line in Figure 9.2(a) and thereby predict the strength of the next cycle.

Now we can do the opposite thing. We can produce a plot in which the horizontal axis corresponds to the strength R_{max} of the cycle n and the vertical axis corresponds to the polar field at the end of the cycle n. Such a plot is shown in Figure 9.2(b). Now we have four data points which we find scattered all over and not lying on a straight line. We say that there is a lack of correlation between the strength of a cycle and the polar field at the end of it. What do all these things mean? Remember the central dogma of the solar dynamo theory that the sunspot cycle is produced by an oscillation between the toroidal and the poloidal magnetic fields (Section 2.5). In Figure 9.3, we indicate an alternation of the toroidal and the poloidal magnetic fields like what happens in the sunspot cycle. Suppose something gives a kind of 'random kick' to the system at the times shown in Figure 9.3. This would provide an explanation of Figures 9.2(a) and 9.2(b). We expect the poloidal field to be correlated with the toroidal field coming after that, as there is no random kick between them. Since the polar field is a manifestation of the poloidal field and the peak sunspot number is an indication of the strength of the toroidal field, this explains the correlation seen in

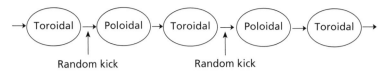

Figure 9.3 A schematic representation of the oscillation between the toroidal and the poloidal magnetic fields, indicating the epochs when the solar dynamo gets random kicks.

Figure 9.2(a). On the other hand, there should not be a significant correlation between the toroidal field and the poloidal field coming after that, because of the random kick between them shown in Figure 9.3. This is why we see a lack of correlation in Figure 9.2(b). The idea encapsulated in Figure 9.3 may seem to follow very naturally from the correlation plots in Figure 9.2. Believe it or not, nobody thought of this idea until it was given in 2007 by me working with two of my PhD students, Piyali Chatterjee and Jie Jiang. The crucial question now is, what provides the random kicks indicated in Figure 9.3? We figured that out too and shall come to a discussion of it a little later.

Another look at Figure 2.11 will convince you that the polar field during the sunspot minimum around 2008 was very weak. If we trust the correlation in Figure 9.2(a), then we would expect the next sunspot cycle to be rather weak. Scientists like Kenneth Schatten and Leif Svalgaard made such a prediction around 2005 when it was becoming clear that the polar field was going to be rather weak during the sunspot minimum. However, since such a prediction is based on very limited statistics, one wonders if we should have too much faith in such a prediction.[2] To give you an example, trials for new drugs are often based on statistical correlation studies. No drug approval agency would approve a new drug based on such scanty statistics on which this sunspot cycle prediction was based. One very pressing question was whether the theoretical dynamo models of the sunspot cycle could throw any light on this issue and could make a theoretical prediction of the strength of the upcoming cycle 24. Around the time of the sunspot minimum between cycles 22 and 23, the flux transport dynamo model was just being developed. The landmark papers by Choudhuri, Schüssler and Dikpati and by Durney appeared in the year 1995. At that time, the solar dynamo models were still too primitive to make any meaningful prediction of the next cycle 23. The sunspot minimum at the end of cycle 23 was the first sunspot minimum when we had reasonably sophisticated models of the solar dynamo. Whether these models could predict the upcoming cycle 24 became a challenging test for these young theoretical solar dynamo models.

Mausumi Dikpati, who was my PhD student at the time we were developing the flux transport dynamo model, had been working in the High Altitude Observatory in Boulder since the completion of her PhD. She teamed up with Peter Gilman to make the first theoretical prediction of cycle 24 based on their dynamo model.[3] In 2006 they predicted that cycle 24 would be the strongest cycle in the last half century— the complete opposite of what Schatten and Svalgaard had predicted on the basis of the weakness of the polar field. If Dikpati and Gilman came out right, then the next data point in Figure 9.2(a) would be way off the straight line. As I went through the two papers of Dikpati and Gilman, it became clear to me that their methodology could not possibly be correct. As I indicated in Figure 9.3, there has to be some kind of a random kick when the toroidal field is getting transformed into the poloidal field, in order to explain the lack of correlation seen in Figure 9.2(b). Dikpati and Gilman had assumed the transformation from the toroidal field to the poloidal field to be a completely regular process. This would not explain the lack of correlation in Figure 9.2(b). There were also some other problems with their model, which I shall discuss later.

Soon after the Dikpati–Gilman prediction was put forth, in a letter to the journal *Nature*, Steven Tobias, David Hughes and Nigel Weiss wrote: 'Any predictions made with such models should be treated with extreme caution (or perhaps disregarded), as they lack solid physical underpinnings.'[4] In spite of my own reservations about the Dikpati–Gilman work, I could not agree with the viewpoint of these critics also. Perhaps I should say a few words here about the viewpoint which these critics have espoused in several of their papers. Many important laws of physics happen to be what are called 'linear' laws. For example, if an electric charge is doubled, we find that the electric field produced by it at a point in space is also doubled. This is an example of a linear law. On the other hand, on doubling the electric charge, if we had found that the electric field became three or four times, it would have been an example of a nonlinear law. It has been known for a few decades that physical systems governed by nonlinear laws often show complicated chaotic behaviour. Nigel Weiss and his collaborators have been arguing over the years that the irregularities of the sunspot cycle are merely such chaotic behaviour arising out of the nonlinearities of the dynamo equation. The problem with this approach is that not all nonlinearities give rise to chaotic behaviour. The simplest kinds of nonlinearities one expects in the dynamo process would make the sunspot cycles more stable rather than producing chaotic behaviour. To get chaotic behaviour out of the solar dynamo, one has to make certain assumptions about the nature of nonlinearities which do not seem to have much support from observational data. So I have advocated the viewpoint that the solar dynamo must be getting some random kicks (as shown in Figure 9.3) to cause the irregularities. I shall very soon come to a discussion of the physical mechanism that produces these random kicks. While discussing magnetoconvection, I mentioned that I regard Nigel Weiss as my second guru (Section 5.2). However, my enormous respect for Nigel has never stopped me from getting into very lively debates with him whenever we have met. Of late, we have been finding that our views are not so divergent now. Nigel now agrees that random kicks play some role in producing irregularities in sunspot cycles, although he still thinks that nonlinearities are the main cause. On the other hand, while I hold the viewpoint that random kicks are the primary cause of sunspot cycle irregularities, I agree that certain patterns in the observational data can be explained best by invoking nonlinearities. Presumably the last word has not yet been said on this subject.

According to the viewpoint of Tobias, Hughes and Weiss, any predictions from theoretical solar dynamo models are impossible beyond very short time ranges. If that were the case, then we would have no explanation for the correlation seen in Figure 9.2(a), just as the Dikpati–Gilman model would leave us without any explanation for the lack of correlation seen in Figure 9.2(b). So I thought that I needed to look at this problem more carefully. I was reasonably sure that the Dikpati–Gilman prediction for cycle 24 was completely wrong. When it would be clear in a few years that cycle 24 was not at all like what the Dikpati–Gilman model had predicted, I was worried that people would lose faith in the whole subject of flux transport solar dynamo. So I felt that somebody should make a theoretical prediction based on more sound scientific arguments. And this had to be done fairly quickly, since the sun was on its way to the next cycle and it would no longer make sense to come up with a prediction after we had started getting the first indications about the nature of cycle 24.

9.4 Gearing Up for the Challenge

Now I should give you, dear reader, a fair warning. During the last few years, it is primarily my group which has worked on developing theories of the irregularities of the sunspot cycle. The remainder of this chapter will mainly present recent results from our group obtained within the last few years. It still remains to be seen whether our ideas are accepted by other research groups and are supported by their independent research. While I am emotionally attached to many of the things I am going to write about, I am fully aware that not all of my ideas will prove correct (hopefully many will). So, read the remainder of this chapter at your own risk. Not everything I write may turn out to be correct.

Towards the end of 2005, I had received an e-mail from my good friend Jingxiu Wang, a senior Chinese solar physicist who had his years of youth disrupted by the Cultural Revolution and then played a key role in building up a tradition of solar research as China came out of the throes of the Cultural Revolution. Jingxiu told me that he and some of his colleagues were convinced that solar dynamo theory was going to be an increasingly important subject and they were concerned that nobody in the whole of China had any expertise in this subject. A young girl, Jie Jiang, who had come out of university with a brilliant academic record and had just joined Jingxiu's group to do a PhD, was willing to learn solar dynamo theory and to work in this field. But she was having difficulties because nobody in China was able to give her guidance. Would I be willing to supervise her in solar dynamo research? Jie had already done some calculations on the solar dynamo and had prepared a preliminary manuscript, which was sent to me. Although the poor English made it difficult to read the manuscript, I went through it. I did not find the problem discussed in the manuscript particularly interesting. Obviously Jie did not know how to choose an interesting research problem. However, I was struck by the methodology used in the calculations. As I finished the manuscript, I knew that only a student with an extraordinary command over mathematical physics could do these calculations. I told Jingxiu that I was willing to supervise Jie. Jingxiu invited me to come to Beijing as visiting professor for several months. I told Jingxiu that, while I would love to spend some time in Beijing, my various commitments made it very difficult to spend more than two months there. It was decided that I would spend two months in the summer of 2006 in Beijing.

Shortly before I was going to leave for Beijing, I read the two Dikpati–Gilman papers on the prediction of cycle 24. As I have already mentioned, I was rather disturbed by these papers and started wondering how one could make a more scientifically sound prediction for cycle 24. In a few days, I started having some ideas on how to approach the problem. Piyali Chatterjee was my PhD student at that time. I told her the following, which I also told Jie Jiang on my arrival in Beijing: 'I have some ideas how to make a more scientifically reliable prediction for cycle 24 than what Dikpati and Gilman have done. I shall be happy if you join me in this project, although I have some doubts whether it will be a good idea for a PhD student to be involved in this project. It is certainly a high-risk project. If we are able to come up with a prediction based on my ideas, it will be about three or four years from now when people will have the first inkling whether our prediction is going to be correct or not. That is exactly the time when you may be looking for a regular academic job. If our prediction turns out to be correct, then that will be a great boost in your job search process. On the other hand, if our prediction fails, nobody may want to offer you a job. Think very carefully whether you want to work on this project and let me know.' Both Piyali and Jie told me that they found my ideas on this subject quite compelling and were willing to take the risk of working on this project.

Although Jie's English was far from fluent at that time, that did not seriously come in the way of our working together. Sometimes when I said something complicated, Jie would request me to repeat it more slowly two or three times. She would also always carry a notebook during scientific discussions. If nothing else worked, then I would have to write down what I was saying. Jie had no problem in reading written English and my bad handwriting also did not discourage her. In 2006 Beijing was still a magical city with traditional neighbourhoods where one could see the traditional Chinese way of life (when I visited Beijing again six years later after the Olympic Games, I was disappointed to see that Beijing was looking too much like other big capital cities of the world and was losing its distinctive character). Although I never missed making sightseeing trips on weekends, on the weekdays we almost worked round the clock during the two months I was in Beijing. In the morning, we would decide which calculation Jie would do that day and usually she managed to finish that by the evening. We were also in constant e-mail contact with Piyali. Although we made very good progress when I was in Beijing, some more things needed to be done by the three of us together to finalize our results. Jie planned a trip to India to finish the work with me and Piyali. Then we ran into an unexpected difficulty. About half a century ago, India had a border war with China. So Indian visa applications from Chinese citizens still undergo a thorough scrutiny. I was surprised to receive a telephone call from the visa officer in the Indian Embassy in Beijing who was handling Jie's visa application. He told me that he himself was fully convinced that this was a genuine case, but he did not have the authority to issue the visa until he had clearance from New Delhi. If I knew any influential person in New Delhi, he suggested that I might ask him or her for help. It was extremely nice of the visa officer to call me. However, I realized to my dismay that I did not know even a single influential person in New Delhi! When I tried to telephone the Ministry of External Affairs myself, the person who picked up the phone was extremely rude to me. I was at my wit's end about what to do. Then somebody told me that Dr. R. Chidambaram, the Principal Scientific Adviser to the Prime Minister, had done his PhD in our department many years ago. He might help me if I wrote to him. I did not know Dr. Chidambaram personally. Still, out of desperation, I wrote to him, introducing myself as a professor in his alma mater and requesting him to help me. Luckily Dr. Chidambaram helped and Jie got the Indian visa.

Before I come to a discussion of our scientific ideas, let me mention that Jie successfully completed her PhD on solar dynamo theory under my guidance and then took up a postdoctoral position in Germany. There she worked with Manfred Schüssler, with whom I had developed the flux transport dynamo model in 1995. Some time ago, Jie has returned to take up a faculty position in Beijing, becoming the first Chinese expert on solar dynamo theory. Since nobody had worked on dynamo theory in India as well before me, may I humbly state that I have been responsible for initiating dynamo research in the two most populous countries of the world—India and China—comprising about one-third of the world's humanity!

9.5 The Die is Cast

Now I come to our work on predicting cycle 24. This work was presented in two papers. We decided to first write a very short paper for one of the high-impact journals which publish only short papers. Then a long paper giving the details of our calculations would appear in a regular journal. We also decided that the order of the authors should be reversed between the two papers, to make it clear to the readers that the contributions of all three of us were equally important. Since the first short paper mainly put forward the central idea which was due to me, I became the first author of this paper. While the order of authors in this first paper was Choudhuri, Chatterjee and Jiang,[5] the second longer paper had Jiang, Chatterjee and Choudhuri listed as authors.[6] Both of these papers appeared in print in 2007. The first paper was selected as 'Editors' Suggestion' in *Physical Review Letters*—one of the world's highest honours for a physics paper. I may mention that we had tried to publish the first paper in two of the world's very high-impact journals: *Nature* and *Science*. Both of these journals expressed unwillingness even to ask for referees' opinions on our paper and rejected it outright.

The first question I bothered about is whether we can predict the next sunspot cycle at all from solar dynamo theory. Not all physical phenomena can be predicted from the laws of physics. When we throw a ball, it is possible to calculate its trajectory from Newton's laws of motion. On the other hand, when we throw a dice, we cannot predict which side of the dice will be up when it hits the floor. Is the progress of the solar dynamo more like the trajectory of a ball or more like the throw of a dice? Obviously the solar dynamo does not consist of a single unified process. It is a complex combination of several processes, as sketched in Figure 7.6. Some of the processes involved in the flux transport solar dynamo seem regular and can be calculated from the basic equations of the subject without any difficulty. For example, the magnetic fields have to be carried by the meridional circulation and the poloidal field has to be stretched by the differential rotation to produce the toroidal field. These seem like processes which can be fully calculated from the basic equations. The other important ingredient of the flux transport solar dynamo is the Babcock–Leighton mechanism for the production of the poloidal field. Let me remind readers that this is a mechanism in which a poloidal magnetic field arises out of the decay of a tilted bipolar sunspot pair, as indicated in Figure 7.1. If the tilt of the sunspot pair is more, then more poloidal field is produced in this process. Now, we have discussed that the tilt of a sunspot pair at a certain latitude is given by Joy's law (Section 2.5). However, Joy's law turns out to be only an average law. Some sunspot pairs seem to have larger tilts than what one expects from Joy's law and some sunspot pairs have smaller tilts. Dana Longcope of Montana State University and I developed a theory for this when I was a visiting professor in Montana for a summer. As a magnetic flux tube rises through the convection zone due to magnetic buoyancy, the Coriolis force acting on it produces the tilt giving rise to Joy's law, as explained by the theoretical model of D'Silva and Choudhuri (Section 5.7). If the convection zone were a completely quiet place, then this certainly would have been the whole story. But we know that there is turbulence within the convection zone and this disturbs the rising flux tubes. That is why we find that there is a scatter in the values of tilt angles around Joy's law when the flux tubes finally emerge on the surface. Because of this, there is an inherent randomness in the Babcock–Leighton mechanism. I arrived at the conclusion that this is likely to be the main source of irregularities in the sunspot cycle.

If we are correct in identifying the Babcock–Leighton mechanism as the main source of randomness in the solar dynamo, then certainly the solar dynamo would be getting some random kicks during the transformation of the toroidal field to the poloidal field which is caused by the Babcock–Leighton mechanism. Note that in Figure 9.3 we precisely identified this phase as the time when the solar dynamo gets random kicks (on the basis of the data plotted in Figure 9.2). Thus, our identifying the Babcock–Leighton mechanism as the main source of randomness in the solar dynamo is consistent with observational data. Now let me discuss how the correlation between the polar field at the sunspot minimum and the strength of the next cycle seen in Figure 9.2(a) arises. Figure 9.4 is taken from the paper by Jiang, Chatterjee and Choudhuri, where an explanation of this correlation was first given. Suppose the poloidal field produced in a sunspot cycle is stronger than average, due to the fluctuations in the Babcock–Leighton mechanism. This will happen if many of the sunspot pairs in that cycle had tilts somewhat on the higher side because of statistical fluctuations. In Figure 9.4 the mid-latitude region near the surface where a lot of poloidal field is expected to be produced is indicated by C. We are considering a situation that the poloidal field produced at C in a sunspot cycle is stronger than the average. Now two things will happen. The poleward meridional circulation near the surface (see Figure 7.6) will carry this strong poloidal field to the pole P to produce a stronger than average polar field at the end of the cycle. We have also discussed turbulent diffusion in the convection zone (Section 6.1). Because of this turbulent diffusion, the poloidal field will try to diffuse from C to the tachocline T,

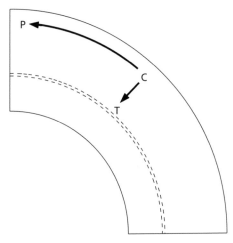

Figure 9.4 A sketch indicating how the correlation between the polar field at the sunspot minimum and the strength of the next cycle arises. Taken from Jiang, Chatterjee and Choudhuri 2007.

where it can act as the seed for the next cycle because the differential rotation there will act on it to produce the toroidal field of the next cycle. If the poloidal field at C is stronger than average to begin with, then we expect the seed for the next cycle to be strong and the next cycle will become stronger than the average. Thus we would have a situation in which the polar field at the end of a cycle is stronger than the average and it is followed by a next cycle which is also stronger than the average. On the other hand, if the randomness in the Babcock–Leighton mechanism makes the poloidal field produced at C weaker than the average, then by similar arguments the polar field at the end of the cycle will be weaker and the next cycle also will be weaker because of the weaker seed giving rise to the cycle. We believe that this is how the correlation seen in Figure 9.2(a) arises.

In order for the correlation to arise in this way, the poloidal field has to diffuse from C to T in a few years in order to act as a seed for the next cycle. Whether this is possible depends on the value of the turbulent diffusion used in the dynamo model. From the nature of turbulence in the convection zone, one can estimate a value of the coefficient of turbulent diffusion. We have used such a value. Many authors before and after us have used such a value of turbulent diffusion which follows from rather simple considerations (called mixing length arguments). In our model with such a turbulent diffusion, we find that the poloidal field takes about 5 years to diffuse from C to T. Our model gives a beautiful correlation between the polar field at the end of a cycle and the strength of the next cycle. Very surprisingly, Dikpati and Gilman chose a value of turbulent diffusion about 50 times smaller than the value which follows from very simple estimates. So, in the Dikpati–Gilman model, the diffusion time is 50 times larger. The poloidal field will take 250 years to diffuse from C to T in their model. Their model does not give any correlation between the polar field at the end of a cycle and the strength of the next cycle, as seen in Figure 9.2(a). I repeat again that the statistical significance of the correlation seen in Figure 9.2(a) is not high. You can take the point of view that the three data points in Figure 9.2(a) happen to lie accidentally near a straight line. As more data points accumulate over the next few solar cycles, we may find them scattered all over Figure 9.2(a) and the correlation we see now may disappear. We certainly cannot completely rule out such a scenario at present. However, if you think that Figure 9.2(a) hints at a true correlation, then only our model can explain this and the Dikpati–Gilman model does not give this correlation. The Dikpati–Gilman model could predict a strong cycle 24 in spite of the weak polar field at the end of cycle 23 only because these are not correlated in this model.

If we run our dynamo code without doing anything special, we get completely periodic cycles. Figure 7.8 showed a typical periodic output from our code. Only when the random kicks indicated in Figure 9.3 are included, would the different cycles become unequal. If we want to model actual sunspot cycles, then we have to put the actual random kicks in the code. We developed a method to figure out the actual random kicks the sun is giving to the dynamo. Since this is a somewhat technical topic, let me just try to give a rough idea without getting into details. The Babcock–Leighton mechanism produces the poloidal field towards the end of the cycle. From the data of the poloidal field, it is possible to extract some information about the strength of the random kick associated with the Babcock–Leighton mechanism at that particular time. Since we have good data for the poloidal field only since the mid-1970s (shown in Figure 7.5), this procedure is possible only from that time. So we can model actual sunspot cycles only from the mid-1970s onwards. Figure 9.5 taken from the Choudhuri, Chatterjee and Jiang paper published in 2007 shows what we got at that time. The solid

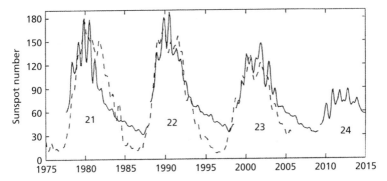

Figure 9.5 A figure from Choudhuri, Chatterjee and Jiang 2007 showing their prediction of cycle 24. The solid line is the theoretical sunspot number calculated from their model, whereas the dashed line is the observed sunspot number.

line is the sunspot number varying with time obtained from our code after incorporating the random kicks on the basis of the poloidal field data. The dashed line is the observational data of sunspot number available up to the time of writing the paper. You can see that the cycles 21, 22 and 23 are modelled reasonably well by our code. Our prediction at that time was that the next cycle 24 would be a very weak cycle. Since our model produces a correlation between the polar field at a sunspot minimum and the strength of the next cycle, and since we have used the fact that the polar field at the end of cycle 23 was weak in order to calculate the random kick at that time, the weakness of cycle 24 is a rather robust result in our model. We could not change it by tweaking things a little bit here and there in the code.

After our paper on predicting cycle 24 came out, some people asked me if we can make a prediction about the next cycle 25 from our model right now. Let me make a comment on this. It is the declining phase of the sunspot cycle when the Babcock–Leighton process creates the poloidal field. Since this process involves randomness, this phase of the sunspot cycle is not predictable. But the rising phase of the sunspot cycle is predictable, because this phase is governed by processes which are regular. Once the unpredictable declining phase of a cycle is nearly over and we have some idea of the random kick that the dynamo has received, we can calculate the predictable rising phase of the next cycle up to the peak of the cycle, but after that our calculations start to become unreliable, as we cannot tell what kind of random kick the dynamo will get during the declining phase. In the future, we may have better dynamo models and better observational data to infer the random kicks. But I do not think that it will ever be possible to make a meaningful prediction of the strength of a cycle more than 7–8 years before its peak.

9.6 The Verdict from the Sun-God

In January 2007 an international conference on 'Challenges for Solar Cycle 24' was organized at the Physical Research Laboratory in Ahmedabad. The organizers wanted to have a session to debate the predictions for cycle 24. They requested Mausumi Dikpati and me to give two successive talks of half-hour duration, presenting our results for cycle 24 prediction. I heard from the organizers that Peter Gilman expressed a wish to give a talk in that session at a rather late date. Normally the organizers would have liked to allot at least half an hour to somebody of Peter's seniority and eminence. But the time table was almost full by that time and the organizers could allot only 15 minutes to Peter. Peter's 15-minute talk was scheduled immediately following the two half-hour talks by Mausumi and me.

Scientists can have differences of scientific opinion and still be very good personal friends. I have good personal friendships with many scientists in our field with whom I disagree on important matters. Unfortunately my personal relationship with Mausumi deteriorated for various reasons after she left Bangalore on completion of her PhD under my supervision. It will not be appropriate for me to give my side of the story here, since readers of this book will then not be able to find out Mausumi's side of the story. So I have decided to restrict myself only to the public aspects of our dispute—in the pages of scientific journals and at international conferences in front of large audiences. At the time of the Ahmedabad conference, my relationship with Mausumi and Peter was at its nadir. I felt rather odd attending a conference where they were present and we were not talking to each other. One of them was my PhD student and the other was the mentor for my postdoctoral research. But such is life. I kept reflecting that the paper I wrote with Peter Gilman in 1987 (Section 5.5) in a way set the ball rolling for my entire scientific career. In that work, we got the first indications that the toroidal field inside the sun may be much stronger than what was hitherto assumed and that the older models of the solar dynamo did not work. This led me to formulation of the flux transport dynamo and the pursuit of its consequences.

As our session began in a sunny morning, Mausumi rose to give the first talk. She made quite a well-organized presentation. But I was rather annoyed that she referred only to her PhD thesis and not to the Choudhuri, Schüssler and Dikpati 1995 paper while summarizing the history of flux transport dynamo theory. Although I found this inappropriate, I did not say anything. Then the second talk was mine. Only the participants of that session can say whether my talk was good or bad. After my talk, Peter came to the stage to make his presentation. The first slide was titled 'Who was the supervisor and who was the student'. Although the solar physics community knew that I was the supervisor of Mausumi, Peter argued for 15 minutes with slide after slide that Mausumi was the real supervisor and I followed her like a student. Perhaps nobody present in the auditorium that day had ever heard such a presentation in a scientific conference. Everybody listened to this completely unprecedented presentation in stunned silence. I knew that I had to say something in response. But I also knew that a long-winded response would not have the desired effect. I had to say something short and crisp. I kept on racking my brain furiously as Peter's talk was coming to an end. Then an idea occurred to me. After Peter finished, I raised my hand to speak. Alexei Pevtsov, an eminent Russian-American solar physicist, was chairing the session. Alexei asked me to speak. I stood up and said: 'I would like to quote a Sanskrit verse which says that a defeat in the hands of your children or your students is like a victory. If Mausumi can become my supervisor, I shall be extremely happy. But wait until you know what the next sunspot cycle is going to be like, before you conclude whether she has become my supervisor.' As soon as I sat down, there was a spontaneous thunderous applause throughout the auditorium. People went on clapping for so long that I wondered if everybody had gone mad. During the tea and lunch breaks of that day, many solar physicists from different countries came over to talk to me, saying that they found me extremely gracious and appreciated my response. Although many years have passed since that international conference, I still often come across solar physicists who tell me that they were present at that conference and vividly remembered the event. Even people who did not attend that conference often tell me that they heard about it from others.

I am always ready for a scientific debate, but I learnt in the years following our predictions how scientific disputes are coloured by sociological factors. Mausumi and Peter were working at the High Altitude Observatory (HAO), which had one of the most powerful solar physics groups in the USA. Some American solar physicists confided to me that they privately agreed that the Dikpati–Gilman work was wrong, but they did not dare to go against HAO publicly. Even in India, such attitudes prevailed. My good friend Debasis Sengupta, a professor in atmospheric and oceanic studies, told me an amusing incident. He was teaching a course on geophysical fluid dynamics. One student asked him if he would cover the theory of the sunspot cycle. Debasis said that the theory of the sunspot cycle required MHD, which was beyond the scope of his course. Then the student said: 'I have heard that one professor of our institute has predicted that the next sunspot cycle will be very weak, whereas his former student has predicted that it will be a very strong cycle.' Debasis asked: 'Who do you think is correct?' The student thought for a while and said: 'I think that the student must be correct. After all, the professor is only a professor in our institute in India. But the student is now a scientist in a top US lab.' At times I certainly did feel like David fighting Goliath.

As months and years rolled by, the sun majestically marched into cycle 24. At the time this book is going to press, we seem to be close to the peak of the cycle. What is the current status of the predictions? Figure 9.6 shows a plot of the sunspot number up to the present time. The small circle on the horizontal time axis indicates the time when the two theoretical predictions were made, whereas the two stars show where the peak of cycle 24 should be according to the two predictions—the upper star corresponding to the Dikpati–Gilman prediction and the lower star corresponding to our prediction. I have deliberately used the monthly sunspot number in this plot, which makes the plot jump up and down. Only when we average over a few months, do we get a smooth plot. I suppose that a lengthy commentary on Figure 9.6 is superfluous. Our prediction seems like a bull's eye which the sun is trying to hit, although the final smoothed curve will probably be slightly lower than even our prediction. In the history of our subject, this is the first almost successful theoretical prediction from a solar dynamo model of the strength of a sunspot cycle before its onset. It seems that the fourth data point in Figure 9.2(a) will lie pretty close to the straight line. I may mention that, even after it became clear that the

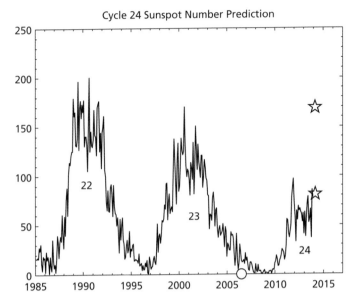

Figure 9.6 The observed monthly sunspot number in recent years, along with the theoretical predictions. The small circle on the time axis indicates the time when the predictions were made, whereas the stars indicate the positions where the peak of cycle 24 should be according to the predictions of Dikpati and Gilman (upper star) and Choudhuri, Chatterjee and Jiang (lower star).

Dikpati–Gilman prediction was wrong, for some time their papers continued to be cited more than our papers. After all, they were in HAO, the Mecca of solar physics, whereas we were toiling in remote Asia.

In science there is a curious asymmetry about the way scientific theories are proved and disproved. Even if a single prediction made from a theory turns out to be wrong, you know that the theory cannot be correct. On the other hand, if one prediction from a theory turns out to be correct, then you wonder if the next predictions will also be correct. How can we ever be sure if a theory is correct or not? This is the celebrated problem of induction in philosophy. Some of the greatest philosophers such as David Hume and Immanuel Kant struggled with this problem. In the twentieth century, Karl Popper claimed to solve this problem. According to him, scientific theories can only be falsified. As long as a theory is not falsified, we have to take it as a correct working theory. According to this Popperian criterion, the Dikpati–Gilman model has been falsified. Since our model passed its first test with flying colours, we have to take it as a correct working model at present. However, one can and should raise the following question: if our methodology is used again in future to predict cycle 25 and still later cycles, will the success it had in predicting cycle 24 be repeated? Nobody can give a definitive answer to this question right now. Let me just put forth my guess.

When we were working on our sunspot cycle prediction based on the idea that the irregularities in the sunspot cycle are caused by fluctuations in the Babcock–Leighton mechanism, we were not aware of another possible source of irregularities in the sunspot cycle. The meridional circulation of the sun acts like a conveyor belt making the sunspot cycle go, as indicated in Figure 7.6. In most of the flux transport dynamo calculations till a few years ago, this conveyor belt had been assumed to be steady. Now some of us think that this conveyor belt can have occasional sudden jerks and can throw things out of gear. We shall discuss these jerks in the next section. We are lucky that there has not been a sudden jerk in the conveyor belt between 2006 when we made our prediction and the present time. That is the reason why our prediction has been so successful. If our methodology is again used during a future sunspot minimum to predict the next sunspot cycle after it, I expect the prediction to be successful again if there is no sudden jerk in the conveyor belt during the intervening period. But a sudden big jerk in the conveyor belt during the intervening period will probably make predictions go haywire.

9.7 Jerks in the Conveyor Belt

The conveyor belt—the meridional circulation inside the sun's convection zone—certainly controls the dynamo. If the conveyor belt becomes faster, then the dynamo will also operate faster, making the duration (i.e. the period) of the cycle shorter. On the other hand, a slowing down of the conveyor belt will make the cycle longer. This important result was discovered by Mausumi Dikpati and Paul Charbonneau in 1999.[7] If the speed of the meridional circulation suddenly changes, that will be reflected in a change in the period of sunspot cycles.

We have actual measurements of the meridional circulation only for the last few years. However, we have data about durations of

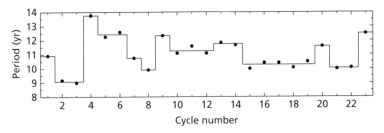

Figure 9.7 Durations of the last few sunspot cycles plotted against the cycle number. The straight lines are to guide the eye to discern patterns in the variations of cycle durations.

sunspot cycles for about three centuries, which give us some indication of whether the meridional circulation had sometimes been faster or slower during this time. Figure 9.7 plots durations of successive cycles against the cycle number. You can see that cycles 9–14 had durations on the longer side, implying that the meridional circulation was probably slower during these cycles. Then these cycles were followed by cycles 15–19 having shorter durations, when the meridional circulation must have been faster. From such indirect means, we conclude that there have been jerks in the conveyor belt—in other words, there have been sudden changes in the speed of the meridional circulation. How will these jerks affect the dynamo? Before my student Bidya Karak and I investigated this question during the last three or four years, nothing was known about this subject.

Suppose the meridional circulation has slowed down. We know that the durations of the sunspot cycles will now be longer. Will there be any other effects? I have pointed out that the turbulent diffusion in our model is about 50 times stronger than that in the Dikpati–Gilman model. The strong diffusion in our model will have more time to act if the duration of the cycles is longer and will make the cycles weaker. Hence we expect that the longer cycles will be weaker and the shorter cycles will be stronger. Exactly the opposite will be true for the Dikpati–Gilman model. In that model, the diffusion is not very effective. If the duration of the cycle is longer due to slower meridional circulation, then differential rotation will have more time to produce more toroidal field and the cycle will be stronger. In the Dikpati–Gilman model, we would expect the longer cycles to be stronger and the shorter

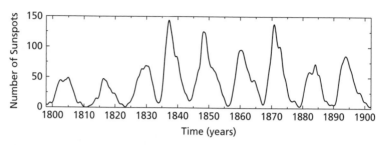

Figure 9.8 The sunspot number plotted against time for some cycles during the nineteenth century when different cycles varied a lot in duration and strength.

cycles to be weaker. Given these two opposite but clear predictions from these two competing models, are there observational data to determine which is the actual situation? Figure 9.8 shows some cycles from the nineteenth century when there were lots of variations from cycle to cycle. If you look carefully, you will agree that there is a tendency for longer cycles to be weaker and shorter cycles to be stronger, providing another strong support to our model against the Dikpati–Gilman model. In fact, Max Waldmeier had already noted in 1935 that stronger cycles seemed to rise faster and weaker cycles seemed to rise more slowly. This is called the *Waldmeier effect*. This is a direct consequence of the fact that stronger cycles are shorter and weaker cycles are longer. In a paper in 2011, Bidya Karak and I provided the first theoretical explanation of the Waldmeier effect.[8] Only our dynamo model can explain this effect. The Dikpati–Gilman model with low diffusion gives exactly the opposite of the Waldmeier effect. This is certainly another feather in the cap of our dynamo model.

We thus conclude that there are two sources for producing irregularities in the sunspot cycle—randomness in the Babcock–Leighton mechanism and fluctuations in the meridional circulation. The most extreme examples of sunspot cycle irregularities are the grand minima like the Maunder minimum, when sunspots practically disappeared for many years. Presumably such grand minima result from a combination of these two sources of irregularities. Bidya and I have been investigating whether we can model grand minima in this way. I have already mentioned the result that there is evidence for 27 grand minima in the last 11,000 years. Given the many uncertainties in our model, we would

Figure 9.9 The author with some younger persons who had worked with him; photograph taken during the first Asia-Pacific Solar Physics Meeting held in Bangalore in 2011. From left to right: Arnab Choudhuri, Dibyendu Nandy, Piyali Chatterjee, Dipankar Banerjee, Jie Jiang, Bidya Karak.

have been happy if our model gave either 10 grand minima or 100 grand minima in 11,000 years. To our great and pleasant surprise, our simulations gave about 24–30 grand minima within 11,000 years in several computer runs.[9] The Choudhuri and Karak 2012 paper also had the good fortune of being selected as 'Editors' Suggestion' in *Physical Review Letters*—an honour which our previous 2007 paper on the prediction of cycle 24 had as well.

A reader who has had the patience to come this far will have developed a good idea of where our field stands now—although without the accompanying mathematical equations and technical details. After developing an understanding of how the sunspot cycle is produced, the thrust of solar dynamo research in the last few years has been to understand the irregularities of the sunspot cycle. The research done in our group certainly appears compelling to me. But I no doubt have a vested interest and am perhaps not an impartial judge. It remains to be seen whether the solar physics community at large will accept some of our more recent ideas.

9.8 A Human Tailpiece

When I was close to finishing this book, a chest X-ray and a CT scan showed a patch on my left lung. Initially doctors suspected it to be lung cancer and I had an extremely interesting week when I was under the impression that I had barely a few more months to live. A difficult biopsy finally established that what I have got is Hodgkin's lymphoma, a cancer of the lymph node which is usually curable with modern medicine. As this book goes to press, I am undergoing regular doses of chemotherapy. On hearing of my illness, both Mausumi Dikpati and Peter Gilman sent me e-mails wishing me a speedy recovery. I was very touched by their e-mails. The e-mail from Peter was the first communication I had from him after the fateful Ahmedabad meeting in 2007. Often, at times of difficulty, we discover our lost friendships. I am glad that I am able to draw a curtain on our unsavoury controversy over cycle 24 prediction with a note of human warmth.

10

Epilogue: Dynamos are Forever

George Ellery Hale's 1908 discovery of magnetic fields in sunspots certainly ushered in a new era in astronomy. Throughout the twentieth century, astronomers came up with many ingenious techniques to look for magnetic fields in distant astronomical systems. All these techniques cannot be explained in a popular science book like this. I can only mention the final outcome. Magnetic fields seem to be all around in the astronomical universe. It is difficult to find a large body of plasma of astronomical size which is completely devoid of magnetic fields. The dynamo process must be a universal process in the astronomical universe.

Our sun's magnetism is certainly not something special to write home about. If our sun were in the position of the nearest star Alpha Centauri, it would have been very difficult to infer with our present-day techniques that the sun has sunspots with a periodic cycle of 11 years. Astronomers are able to study magnetic fields of mainly those stars in which magnetic fields produce much more dramatic effects—spots much larger than sunspots, cycles much more pronounced, flares compared to which the strongest solar flares appear timid. And there are numerous such stars. However, for magnetic fields of no other astronomical system, do we have as rich a treasure trove of observational data as we have for the sun, to guide us in building theoretical models. First of all, we can see sunspots on the face of the sun—sometimes even with the naked eye. Even the most powerful telescopes do not allow us to directly see starspots on the face of any star. We have to study starspots only by indirect methods. There are really ingenious and marvellous methods which astronomers have devised in order to study starspots. But these can never really substitute for the thrill of seeing a spot with your own eyes. Also, we have lots of data on how the sun's magnetic field has behaved with time. Even before it was realized that sunspots are manifestations of the sun's magnetic field, astronomers had been monitoring sunspot cycles. Now C-14 and Be-10 are pushing our records back by many millennia, to even before the dawn of human civilization.

Our understanding of the sun's dynamo gives us vital clues towards understanding the dynamo process in different astronomical systems. The main purpose of this book has been to tell you about sunspots, their cycle and how they affect us. However, I do not want to say a final good-bye without giving you a glimpse of the wide vista that opens before us as our understanding of the dynamo process deepens from our study of the sun's dynamo. This final Epilogue will be a short rambling tour through the world of astronomical magnetic fields and astronomical dynamos.

The spectrum of the sun has absorption lines of calcium called the H and K lines. However, when one takes the spectrum of a region around a large sunspot with strong magnetic fields, one often finds emission at these spectral lines instead of absorption. I shall not try to give a theoretical explanation of this, which is somewhat complicated. Just remember the following: enhanced emission in the H and K spectral lines of calcium instead of absorption indicates some magnetic activity in the region from where the light comes. G. Eberhard and Karl Schwarzschild reported in 1913 that they found this kind of emission in calcium H and K lines in the spectra of several stars. It became clear that these stars must have really strong magnetic fields at their surface. If these magnetic fields varied in a cycle like the sunspot cycle, then we would expect a variation of the emission in the calcium H and K lines. However, if these cycles have periods of a few years like the 11-year cycle of the sun, then we have to monitor the calcium H and K emissions from these stars for many years to discover their cycles. In the 1960s Olin Wilson initiated a programme at the Mount Wilson Observatory to monitor calcium H and K emissions from several stars over many years. With data collected over many years, some of the monitored stars seem to have fairly regular cycles.[1] Other stars do not show much variation with time. Perhaps most intriguing are the stars which seem to be entering grand minima like the Maunder minimum or are coming out of such grand minima. It is fascinating to look at the time evolution plots of these stars and to compare them with the plot of sunspot number. Astronomers have succeeded in measuring the rotation periods of many of these stars. It seems that the calcium H and K emissions tend to be stronger from stars which are rotating faster. This is exactly what we would expect from very simple considerations. We have seen that the sun's rotation plays a crucial role in the dynamo process. Stars that rotate faster are expected to have stronger dynamos giving rise to more magnetic fields and more emissions in calcium H and K lines.

I have already mentioned that some stars have spots of a much more gigantic size than sunspots. There is a clever method of mapping the large starspots on a star's surface (called Doppler imaging), which would be somewhat too technical to explain in this book. It seems that some of the rapidly rotating stars get large starspots in the polar region. I remind readers about the discussion of magnetic buoyancy and the Coriolis force in Chapter 5. The interplay between these two forces was first studied in 1987 by me and Peter Gilman (Section 5.5). The magnetic buoyancy has to overcome the Coriolis force for a flux tube to come out radially. On the other hand, if the star is rotating very fast and the Coriolis force is strong, the rising flux tube would be diverted to the polar region, as indicated by the line with dots in Figure 5.5. In 1992 Manfred Schüssler and Sami Solanki simply extrapolated the results of the Choudhuri and Gilman 1987 paper to explain how polar spots form in rapidly rotating stars.[2] When I met Sami Solanki at a conference, I told him: 'Before I read your paper, I did not know about these observations of polar starspots in rapidly rotating stars. Had I known about these observations, then I could have worked out all the things in your paper in two or three days.' Sami laughed and said: 'I know it. We are lucky that you did not know about these observations. That is why we got the chance of writing this interesting little paper.'

The study of planetary magnetic fields has also blossomed in the last few decades. Before the dawn of the Space Age, the earth was the only planet whose magnetic field scientists could investigate. Now, spacecraft missions flying near other planets of the solar system have given us considerable information about magnetic fields of other planets. Theoreticians have also been busy making models of planetary dynamos. Unlike the sun which has an oscillating magnetic field, the magnetic field of the earth appears to be approximately static. However, as a result of the pioneering research done by Bernard Brunhes and Motonori Matuyama in the early years of the twentieth century, we now know that the earth's magnetic field has sometimes reversed in the past—the north magnetic pole becoming the south magnetic pole and vice versa. Many rocks have magnetic particles in them. The orientations of these magnetic particles give us the direction of the earth's magnetic field at the time when the rock formed. So, if we can date when a rock formed, we can figure out the direction of the earth's magnetic field at that time. Geological records now reveal many reversals of the earth's magnetic field in the past. As far as we can tell, there is no particular pattern or periodicity in these reversals. They seem like random events. The last geomagnetic reversal occurred about 780,000 years ago.

As we have seen, the differential rotation of the sun plays a crucial role in the solar dynamo. The earth does not have any significant differential rotation. We, of course, know that the earth's surface is solid and the angular velocity of rotation is constant over it. The constancy of angular velocity approximately holds for the earth's interior as well. It seems that the sun's differential rotation is very important in making the dynamo oscillatory. In the absence of differential rotation, the dynamo tends to be non-oscillatory, as in the case of the earth. In Parker's original idea of solar dynamo as encapsulated in Figure 6.4, helical turbulence would twist the toroidal field to produce the poloidal field, whereas it is the differential rotation which would lead from the poloidal field to the toroidal field. Since differential rotation is unimportant for the earth, the helical turbulence in the earth's core has also to do the job of twisting the poloidal field to produce the toroidal field. In other words, for the case of the earth's dynamo, we have to put 'Helical turbulence' in both the rectangular boxes above and below in Figure 6.4. As we have discussed in Chapter 6, Parker in 1955 and then Steenbeck, Krause and Rädler in 1966 developed a method of averaging over turbulence. Such an approach in which we average over turbulence is called the *mean field* dynamo theory. Nearly all the results of solar dynamo we have discussed in this book are from such a mean field theory. However, since turbulence arises out of the basic equations of fluid mechanics, one could also generate turbulence in computer models by solving the equations of fluid mechanics (primarily the Navier–Stokes equation introduced in Section 6.1). This approach is called *direct numerical simulation* (DNS) and requires very large computers.

In 1995 Gary Glatzmaier and Paul Roberts used the supercomputer of the Los Alamos National Laboratory—one of the world's largest computers at that time—to do a DNS of the earth's dynamo.[3] After running for 2000 CPU hours on this supercomputer, their code suddenly showed a reversal of the earth's magnetic field. This was the first success in theoretically producing a geomagnetic reversal from a dynamo model, and it threw a lot of light on how such reversals take place. During the reversal, the earth's magnetic field presumably passes through a phase when it becomes very weak. We have pointed out that the earth's magnetic field produces the magnetosphere, which is like a shield protecting us from the excesses of solar disturbance. During a geomagnetic reversal, does the magnetosphere disappear altogether? Will that allow the solar wind and the coronal mass ejections to almost reach the earth? What will the consequences of this be? We can only speculate. There have been conjectures that even some of the mass extinctions of species in the past may be connected with geomagnetic reversals.

The sun's oscillatory magnetic field is much more complicated than the earth's nearly static magnetic field. Doing a full DNS of the solar dynamo is a much more challenging job than doing a DNS of the earth's dynamo. Only recently, some groups like those of Allan Sacha Brun and Paul Charbonneau have made some progress in doing DNS of the solar dynamo. Still the initial results are of a rather exploratory nature and are very far from providing detailed models of different aspects of the sunspot cycle. We still have to rely on the mean field models when we want to make detailed matches between observations and theory. However, it is quite possible that major breakthroughs will take place in the DNS of the solar dynamo within the next few years. If I were about 10 or 15 years younger, I might have wanted to get into this field. Cutting-edge frontier research is very much a young people's game. Now I have probably reached that stage of life when I can only watch younger scientists achieve spectacular new things that I can no longer do.

At the end of our long journey, we come to galaxies. It is one of the achievements of twentieth-century astronomy to realize that the galaxies are the building blocks of the universe. You can clearly see a beautiful spiral structure in the image of the galaxy shown in Figure 10.1. Our galaxy is also believed to have such a spiral structure. Astronomers have found that such spiral galaxies tend to have magnetic fields—with magnetic field lines usually running along the spiral arms of the galaxy. Here I shall not discuss the very ingenious method by which William Hiltner in the 1950s first concluded that our galaxy the Milky Way has a magnetic field, since it would require a few pages to explain this method. Let me mention another method by which astronomers can infer the existence of magnetic fields in external galaxies and other astrophysical systems. We have discussed cosmic rays (Section 8.1)—highly accelerated charged particles zooming around through our galaxy. Presumably other galaxies similar to ours will also

Figure 10.1 The famous Whirlpool Galaxy M51 possessing a spiral structure, photographed with the Hubble Space Telescope. Credit: ESA, NASA and Space Telescope Science Institute.

contain such accelerated charged particles and these particles are expected to spiral around the magnetic fields of these galaxies due to the Lorentz force exerted by these magnetic fields. Now, highly accelerated charged particles spiralling around magnetic field lines emit a special kind of radiation known as *synchtrotron radiation*. As soon as astronomers receive synchrotron radiation from an external galaxy or some other astrophysical source, they at once know that magnetic fields must be present at the source of this radiation. How the dynamo process produces the magnetic field of a galaxy is now a very hot research topic. This is a field in which many seminal contributions came from the Soviet Union just before its breakup. The legendary Russian physicist Yakov Zeldovich and some of his students were pioneers in this field. Apart from stars, a spiral galaxy also contains gas in the interstellar space between stars. The turbulence in this gas gives rise to the dynamo process in the galaxy.

People working on the galactic dynamo have to bother about one question which scientists working on stellar or planetary dynamos almost never worry about. We have seen that the dynamo process involves the poloidal and the toroidal magnetic fields sustaining each other. In other words, we need some magnetic fields to begin with. The dynamo process can sustain or amplify magnetic fields, but cannot create them if there are absolutely no initial magnetic fields at all. Systems like our solar system have formed by condensing out of interstellar gas clouds. Since the galactic magnetic field threads through the interstellar gas clouds, this magnetic field can provide the seed magnetic fields on which the stellar or planetary dynamos start to work. However, in order for a galactic dynamo to start, we need some initial magnetic field in the galaxy. The dynamo process could not create the first magnetic fields of the universe. How the first magnetic fields of the universe arose is a question to which we still do not know the definitive answer. There are many theoretical speculations, but we have no universally accepted theory as yet.

An interstellar gas cloud sometimes undergoes a collapse under its own gravity and eventually produces stars. Magnetic fields are believed to play a very important role in this star formation process. Suppose you are standing on a rotating platform with your hands outstretched. If you put your hands down, you will find that you will start rotating faster. This follows from the conservation of angular momentum. Since the disc of a galaxy rotates around its centre, any interstellar gas cloud would have some angular momentum. As it collapses, it will start rotating faster and faster due to the conservation of angular momentum. Eventually the centrifugal force due to this rotation would become so large that it would balance the gravity and the collapse of the gas cloud would come to a halt. We have discussed in Section 8.2 how magnetic fields in the solar wind take away angular momentum from the sun, braking the rotation of the sun. Exactly in the same way, magnetic fields in the interstellar gas cloud are expected to remove angular momentum from the collapsing gas cloud and thereby allow the collapse to proceed. If there were no magnetic fields, then this would not happen and any galaxy would consist of many spinning gas clouds which could not collapse any further to form stars. If this were the case, then our present-day universe would have been a rather dark place because there would be no stars to give out starlight. In such a situation, there would obviously be no such thing as our solar system. A minor consequence of that would have been that I would not be here to write this book and you would not be here to read it. This is truly an existential dilemma. Dear reader, the existences of you and me are made possible courtesy of the astronomical magnetic fields and courtesy of the dynamo process.

Appendices: Technical Details of some Important Topics

These appendices are meant for those readers who would like to see a little bit of technical details for some of the important topics discussed in the book. I assume a college-level knowledge of physics and mathematics (especially a knowledge of vector calculus) for those who wish to read the appendices.

Appendix A. The Ideal Gas Law and Other Equations of State

Let us consider n moles of an ideal gas. Its pressure P, volume V and temperature T (in kelvin) are related by the ideal gas law

$$PV = nRT, \tag{1}$$

where R is the universal gas constant. This law is a combination of the well-known Boyle's law and Charles's law.

The theoretical study of stellar structure made rapid progress when astrophysicists such as Eddington realized that the matter inside ordinary stars like the sun can be regarded as an ideal gas. However, when considering the gas inside stars, it is more convenient to talk in terms of the density ρ rather than the volume V of n moles of the gas. Let us point out how we can change over from V to ρ. The number of gas particles within n moles of the gas is nN_A, since the number of gas particles per mole is the Avogadro number N_A. If μ is the mass per gas particle, then the mass of n moles is $nN_A\mu$ and the density is

$$\rho = \frac{nN_A\mu}{V}. \tag{2}$$

Writing $R = N_A k$ (where k is the Boltzmann constant), (1) and (2) give

$$P = \frac{k}{\mu}\rho T. \tag{3}$$

The ideal gas law in this form is used in the study of stellar structure. Some special care has to be taken when the gas particles are of different kinds and have different masses (electrons, protons, helium nuclei, ...). We do not discuss these details here.

When the nuclear fuel is exhausted, a star collapses to very high density and the ideal gas law no longer holds. If relativistic saturation of particle velocities to the speed of light c is not taken into account, then the pressure P of such high-density 'degenerate' matter is related to the density ρ in the following way:

$$P \propto \rho^{5/3}. \tag{4}$$

On the other hand, on taking account of the relativistic saturation of particle velocities, we get

$$P \propto \rho^{4/3}. \tag{5}$$

Equation (4) would allow white dwarfs of any mass. However, one can derive from (5) that there is a maximum mass limit for white dwarfs (the Chandrasekhar mass limit).

Appendix B. Hydrostatic Equilibrium and the Condition for Convection

For a fluid to be in hydrostatic equilibrium, the following equation has to be satisfied:

$$\nabla P = \rho \mathbf{g}, \tag{6}$$

where $\mathbf{g}$ is the gravitational field at a point where the pressure and the density of the fluid are P and ρ respectively.

Let us consider a spherically symmetric star. Suppose M_r is the mass within the radius r. It is easy to show that the gravitational field at radius r would be produced by only the mass inside r and would be given by

$$\mathbf{g} = -\frac{GM_r}{r^2}\mathbf{e}_r, \tag{7}$$

where G is the gravitational constant and $\mathbf{e}_r$ is the unit vector directed radially outward. In a spherically symmetric situation, ∇P becomes $dP/dr\,\mathbf{e}_r$ so that (6) gives

$$\frac{dP}{dr} = -\frac{GM_r}{r^2}\rho, \tag{8}$$

which is the hydrostatic equilibrium equation for stars and is one of the fundamental equations of stellar structure. Another equation of stellar structure which you may try to derive yourself is

$$\frac{dM_r}{dr} = 4\pi r^2 \rho. \tag{9}$$

I shall not discuss the other equations of stellar structure. Let me just state the Schwarzschild condition that the temperature gradient $|dT/dr|$ has to be larger than a critical value for convection to take place. The Schwarzschild condition for convection to occur is

$$\left|\frac{dT}{dr}\right| > \left(1 - \frac{1}{\gamma}\right)\frac{T}{P}\left|\frac{dP}{dr}\right|, \tag{10}$$

where γ is the adiabatic gas index and the other symbols have the usual meanings.

For a discussion of the stellar structure equations and derivation of the Schwarzschild condition, see Choudhuri, *Astrophysics for Physicists*, pp. 61–73.

Appendix C. The Constancy of Magnetic Flux through a Coil of Zero Resistivity

Suppose Φ_{ext} is the magnetic flux through a coil due to an external magnetic field. If this flux changes, one important law of electromagnetism is that the EMF induced in the coil is given by $-d\Phi_{ext}/dt$. This will induce a current I through the coil, which will give rise to an additional magnetic flux LI through the coil, L being the self-inductance. If R is the resistance of the coil, then the standard LR circuit equation gives us

$$-\frac{d\Phi_{ext}}{dt} = L\frac{dI}{dt} + RI, \tag{11}$$

from which

$$-\frac{d}{dt}(\Phi_{ext} + LI) = RI.$$

If $R = 0$, it at once follows that

$$\Phi_{ext} + LI = \text{constant}. \tag{12}$$

Now, $\Phi_{ext} + LI$ is the total magnetic flux through the coil, which does not change if $R = 0$. When the flux Φ_{ext} due to the external magnetic field changes, a current I is induced in such a way that the flux LI through the coil due to this current exactly compensates the change in Φ_{ext}.

Appendix D. Maxwell's Equations and the Induction Equation of MHD

The electric field $\mathbf{E}$ and the magnetic field $\mathbf{B}$ arise from the charge density ρ and the current density $\mathbf{j}$. The famous Maxwell's equations tell us how all of these are connected. They are

$$\nabla.\mathbf{E} = \frac{\rho}{\epsilon_0}, \tag{13}$$

$$\nabla.\mathbf{B} = 0, \tag{14}$$

$$\nabla \times \mathbf{B} = \mu_0 \left(\mathbf{j} + \epsilon_0 \frac{\partial \mathbf{E}}{\partial t} \right),$$ (15)

$$\nabla \times \mathbf{E} = -\frac{\partial \mathbf{B}}{\partial t},$$ (16)

where ϵ_0 and μ_0 are the two fundamental constants of electromagnetism. The second term on the right-hand side of (15) is known as the displacement current and is the original contribution of Maxwell. It is this term which was crucial for Maxwell's prediction that light is an electromagnetic wave. However, when we consider plasma motions in MHD with speeds small compared to the speed of light c, the displacement current term turns out to be small compared to the other terms in the equation. The approximation of neglecting this term is called the MHD approximation. In this approximation, we replace (15) by the simpler equation

$$\nabla \times \mathbf{B} = \mu_0 \mathbf{j}.$$ (17)

This is known as Ampere's equation and describes how magnetic fields arise out of electric currents when rapid time variations are not important.

Apart from Maxwell's equation, the other fundamental equation of electromagnetism is the Lorentz force equation giving the force $\mathbf{F}$ acting on a charge q moving with speed $\mathbf{v}$ in an electromagnetic field:

$$\mathbf{F} = q(\mathbf{E} + \mathbf{v} \times \mathbf{B}).$$ (18)

Since charged particles moving under electromagnetic forces give rise to the current density $\mathbf{j}$, we expect $\mathbf{j}$ to be given by

$$\mathbf{j} = \sigma(\mathbf{E} + \mathbf{v} \times \mathbf{B}),$$ (19)

which is known as Ohm's law for plasmas, σ being the electrical conductivity.

Now, from (19) we have

$$\mathbf{E} = \frac{\mathbf{j}}{\sigma} - \mathbf{v} \times \mathbf{B}.$$

Substituting for $\mathbf{j}$ from (17), we get

$$\mathbf{E} = \frac{\nabla \times \mathbf{B}}{\mu_0 \sigma} - \mathbf{v} \times \mathbf{B}.$$

We now have to substitute this expression for $\mathbf{E}$ in (16), which is nothing but Faraday's law of electromagnetic induction. This gives

$$\frac{\partial \mathbf{B}}{\partial t} = \nabla \times (\mathbf{v} \times \mathbf{B}) + \frac{1}{\mu_0 \sigma} \nabla^2 \mathbf{B},$$ (20)

if we assume σ to be constant in space. This is the induction equation, which is the central equation in MHD.

Appendix E. The Induction Equation and Flux Freezing

The first term on the right-hand side of (20) implies that the magnetic field in a plasma changes due to motions in the plasma, whereas the second term implies that the magnetic field diffuses due to the resistivity (inverse of σ) of the plasma. We can make an approximate estimate of the relative importance of these two terms in the following way. Let V and B be typical values of the velocity $\mathbf{v}$ and the magnetic field $\mathbf{B}$ inside the plasma, whereas L is the typical length scale over which these quantities change appreciably. The first term on the right-hand side of (20) should be of order VB/L, whereas the second term should be of order $B/\mu_0 \sigma L^2$. The ratio of these two terms is the dimensionless number known as the magnetic Reynolds number

$$\mathcal{R}_{M} = \frac{VB/L}{B/\mu_0 \sigma L^2} = \mu_0 \sigma\, VL. \tag{21}$$

Since $\mathcal{R}_{M}$ is proportional to L, it is larger for systems of larger size. This means that the magnetic Reynolds number $\mathcal{R}_{M}$ for astrophysical systems should be much larger than $\mathcal{R}_{M}$ for typical laboratory systems. For astrophysical systems, usually $\mathcal{R}_{M}$ is much larger than 1, which means that the first term on the right-hand side of (20) is usually much more important for astrophysical systems compared to the second term. While dealing with astrophysical plasmas, if we keep only the first term, then (20) leads to

$$\frac{\partial \mathbf{B}}{\partial t} = \nabla \times (\mathbf{v} \times \mathbf{B}). \tag{22}$$

It is this equation from which Alfvén's theorem of flux freezing follows.

Let me give the mathematical statement of the theorem. At time t_1, let us consider the magnetic flux $\int_{S_1} \mathbf{B}.d\mathbf{S}$ through a surface S_1 in the magnetized fluid. The fluid elements which made up this surface S_1 at time t_1 will all move with time and will make a different surface S_2 at time t_2. Let $\int_{S_2} \mathbf{B}.d\mathbf{S}$ be the magnetic flux through this surface at time t_2. According to Alfvén's theorem of flux freezing, we must have

$$\int_{S_2} \mathbf{B}.d\mathbf{S} = \int_{S_1} \mathbf{B}.d\mathbf{S},$$

if the magnetic field evolves according to (22). For those who are comfortable with vector analysis, the proof of this theorem is not particularly difficult. For a proof of the theorem, see Choudhuri, *Astrophysics for Physicists*, pp. 230–232. From the basic physics point of view, this theorem is the MHD generalization of the constancy of magnetic flux through a coil of resistance zero, discussed in Appendix C.

Appendix F. Magnetic Forces within the Plasma

When we consider electromagnetic forces acting on charge and current distributions instead of a moving charged particle, we have to replace (18) by the following expression of the force **F** per unit volume:

$$\mathbf{F} = \rho \mathbf{E} + \mathbf{j} \times \mathbf{B}. \tag{23}$$

In a typical plasma, the charge density ρ would be close to zero, but the current density **j** can be important. So it is the magnetic part of the force in (23) which is important in MHD and let us focus our attention on only this part:

$$\mathbf{F}_{mag} = \mathbf{j} \times \mathbf{B}.$$

On replacing **j** by making use of (17), we get

$$\mathbf{F}_{mag} = \frac{(\nabla \times \mathbf{B}) \times \mathbf{B}}{\mu_0}. \tag{24}$$

On making use of a standard vector identity, (24) becomes

$$\mathbf{F}_{mag} = \frac{(\mathbf{B}.\nabla)\mathbf{B}}{\mu_0} - \nabla \left(\frac{B^2}{2\mu_0} \right). \tag{25}$$

Motions inside fluids are governed by Euler's equation, the central equation of fluid mechanics. It is

$$\rho \left[\frac{\partial \mathbf{v}}{\partial t} + (\mathbf{v}.\nabla)\mathbf{v} \right] = -\nabla P + \rho \mathbf{g}, \tag{26}$$

where **v** is the velocity of a fluid element at the point where we are considering this equation. The hydrostatic equilibrium equation (6) follows simply on putting $\mathbf{v} = 0$ in (26).

Now we consider a situation in which the gravitational field **g** is unimportant, but the fluid is a magnetized plasma so that we need to add the magnetic force $\mathbf{F}_{mag}$ as given by (25) in (26). On doing this, we get

$$\rho \left[\frac{\partial \mathbf{v}}{\partial t} + (\mathbf{v}.\nabla)\mathbf{v} \right] = -\nabla \left(P + \frac{B^2}{2\mu_0} \right) + \frac{(\mathbf{B}.\nabla)\mathbf{B}}{\mu_0}. \tag{27}$$

This is one of the important equations of MHD showing how motions inside plasmas are affected by magnetic forces. One contribution of the magnetic field is the term $B^2/2\mu_0$ which gets added to the pressure P. This term is exactly like an additional pressure and is called the magnetic pressure. The other term $(\mathbf{B}.\nabla)\mathbf{B}/\mu_0$ acts like a tension force and is called the magnetic tension. To realize that this term is like a tension, you can easily check that this term will be zero if the magnetic field consists of straight field lines in a region. Only when magnetic field lines are bent, this force arises and tries to straighten the field lines so that this tension force gets reduced.

Appendix G. Magnetic Buoyancy

We consider a horizontal magnetic flux tube with magnetic field B inside it and very little magnetic field outside. If P_i is the gas pressure inside the flux tube, it follows from (27) that $P_i + B^2/2\mu_0$ should act like the total pressure inside the flux tube. Since we would expect a pressure balance across the surface of the flux tube (so that the flux tube does not expand or contract), the external gas pressure P_e outside the flux tube (where there is no magnetic pressure) must equal the total pressure inside, i.e.

$$P_e = P_i + \frac{B^2}{2\mu_0}. \tag{28}$$

This clearly implies

$$P_i < P_e, \tag{29}$$

from which it often, though not always, follows that the internal density ρ_i has to be less than ρ_e. For example, if the temperatures inside and outside are the same, then a straightforward application of (3) to (29) implies

$$\rho_i < \rho_e. \tag{30}$$

If (29) holds for a portion of the magnetic flux tube, then that portion will clearly be buoyant due to Archimedes's principle.

Appendix H. Mean Field MHD and Dynamo Theory

The dynamo action takes place in the turbulent convection zone of the sun. In a turbulent situation, we can split quantities like $\mathbf{v}$ and $\mathbf{B}$ into a mean part and the fluctuation around the mean:

$$\mathbf{v} = \bar{\mathbf{v}} + \mathbf{v}', \quad \mathbf{B} = \bar{\mathbf{B}} + \mathbf{B}', \tag{31}$$

where the overline indicates the mean and the prime indicates the fluctuation around the mean. On substituting (31) in (20), we get

$$\frac{\partial \bar{\mathbf{B}}}{\partial t} + \frac{\partial \mathbf{B}'}{\partial t} = \nabla \times (\bar{\mathbf{v}} \times \bar{\mathbf{B}} + \bar{\mathbf{v}} \times \mathbf{B}' + \mathbf{v}' \times \bar{\mathbf{B}} + \mathbf{v}' \times \mathbf{B}') + \frac{1}{\mu_0 \sigma} \nabla^2 (\bar{\mathbf{B}} + \mathbf{B}').$$

We now take average of the whole equation term by term. Keeping in mind that the averages of $\mathbf{v}'$ and $\mathbf{B}'$ are zero by definition, we get

$$\frac{\partial \bar{\mathbf{B}}}{\partial t} = \nabla \times (\bar{\mathbf{v}} \times \bar{\mathbf{B}}) + \nabla \times \mathcal{E} + \frac{1}{\mu_0 \sigma} \nabla^2 \bar{\mathbf{B}}, \tag{32}$$

where

$$\mathcal{E} = \overline{\mathbf{v}' \times \mathbf{B}'} \qquad (33)$$

does not necessarily have to be zero, even though the averages of $\mathbf{v}'$ and $\mathbf{B}'$ individually are zero.

It is this quantity $\mathcal{E}$ given by (33) which is crucial for dynamo action. Steenbeck, Krause and Rädler developed a mean field theory of MHD to evaluate $\mathcal{E}$ and showed that it should have the form

$$\mathcal{E} = \alpha \overline{\mathbf{B}} - \beta \nabla \times \overline{\mathbf{B}}. \qquad (34)$$

It is possible to evaluate α and β from the mean field theory of MHD, though we shall not discuss these details here. For a full discussion, see Choudhuri, *The Physics of Fluids and Plasmas*, pp. 350–355.

On substituting (34) in (32), we finally have

$$\frac{\partial \overline{\mathbf{B}}}{\partial t} = \nabla \times (\overline{\mathbf{v}} \times \overline{\mathbf{B}}) + \nabla \times (\alpha \overline{\mathbf{B}}) + \left[\frac{1}{\mu_0 \sigma} + \beta\right] \nabla^2 \overline{\mathbf{B}}, \qquad (35)$$

which is the celebrated dynamo equation, the central equation of dynamo theory. Parker had derived a form of this equation in his famous 1955 paper by heuristic arguments. It is easily seen that the coefficient β adds to the diffusion. Indeed, β is the coefficient of turbulent diffusion and is usually much larger than $(\mu_0 \sigma)^{-1}$ when turbulence is present, which means that turbulence provides the main diffusion mechanism in the dynamo process.

It is the term $\nabla \times (\alpha \overline{\mathbf{B}})$ in (35) which is at the heart of dynamo theory and makes the form of the dynamo equation (35) different from the induction equation (20). Cowling's theorem, which holds for (20), does not hold for (35). By solving (35) under suitable conditions, it is indeed found that an axisymmetric mean field $\overline{\mathbf{B}}$ can be sustained by fluid motions of which the mean part $\overline{\mathbf{v}}$ is axisymmetric. The equations for the flux transport dynamo shown in Figure 7.7 arise out of (35).

For modelling the sunspot cycle, one requirement is to have a dynamo wave propagating equatorward. In order to have equatorward propagation in the absence of meridional circulation, the condition that has to be satisfied is

$$\alpha \frac{\partial \Omega}{\partial r} < 0 \qquad (36)$$

in the northern hemisphere, Ω being the angular velocity inside the sun. This condition is known as the Parker–Yoshimura sign rule. A derivation of this rule can be found in Choudhuri, *The Physics of Fluids and Plasmas*, Section 16.6.

Notes

Chapter 1: Explosions, Blackouts and Cycles

1 Carrington, R. C., 1859, 'Description of a singular appearance seen in the Sun on September 1, 1859', *Monthly Notices of the Royal Astronomical Society*, **20**, 13–15.

2 Tsurutani, B. T., Gonzalez, W. D., Lakhina, G. S., & Alex, S., 2003, 'The extreme magnetic storm of 1–2 September 1859', *Journal of Geophysical Research*, **108**, 1268 (8 pp).

3 Schwabe, H., 1844, 'Sonnen-Beobachtungen im Jahre 1843', *Astronomische Nachrichten*, **21**, 233–236.

4 Herschel, W., 1795, 'On the nature and construction of the Sun and fixed stars', *Philosophical Transactions of the Royal Society of London*, **85**, 46–72.

5 Hale, G. E., 1908, 'On the probable existence of a magnetic field in sun-spots', *Astrophysical Journal*, **28**, 315–343.

6 Cowling, T. G., 1933, 'The magnetic field of sunspots', *Monthly Notices of the Royal Astronomical Society*, **94**, 39–48.

7 Krause, F., 1993, 'The cosmic dynamo: From $t = -\infty$ to Cowling's theorem. A review on history' in *The Cosmic Dynamo: Proceedings of the 157th Symposium of the International Astronomical Union*, p. 497.

8 Parker, E. N., 1955, 'Hydromagnetic dynamo models', *Astrophysical Journal*, **122**, 293–314.

Chapter 2: The Mysterious Sunspots

1 Vaquero & Vázquez, *The Sun Recorded Through History*, p. 112.

2 Carrington, R. C., 1859, 'On certain phenomena in the motions of solar spots', *Monthly Notices of the Royal Astronomical Society*, **19**, 81–84.

3 Eddy, J. A., 1976, 'The Maunder Minimum', *Science*, **192**, 1189–1202. The quotation from Cassini is on p. 1190.

4 Clerke, *A Popular History of Astronomy during the Nineteenth Century*, p. 161.

5 Carrington, R. C., 1858, 'On the distribution of the solar spots in latitudes since the beginning of the year 1854', *Monthly Notices of the Royal Astronomical Society*, **19**, 1–3.

6 Maunder, E. W., 1904, 'Note on the distribution of sun-spots in heliographic latitude, 1874-1902', *Monthly Notices of the Royal Astronomical Society*, **64**, 747–761.

7 Spörer, G. 1874, *Beobachtungen der Sonnenflecken zu Anclam* (W. Engelmann Liepzig).

8 Clark, *The Sun Kings*, p. 99.

9 Hale, G. E., Ellerman, F., Nicholson, S. B., & Joy, A. H., 1919, 'The magnetic polarity of sun-spots', *Astrophysical Journal*, **49**, 153–178.

10 For those readers who have some mathematical training, let me explain how the toroidal and the poloidal magnetic fields are defined mathematically. In dynamo calculations we often assume the magnetic field to be symmetric around the rotation axis, which is taken as the polar axis for introducing spherical polar coordinates. In these coordinates, the magnetic field can be written as

$$\mathbf{B} = B_r \mathbf{e}_r + B_\theta \mathbf{e}_\theta + B_\phi \mathbf{e}_\phi,$$

where the symbols have the usual meanings which should be clear to anybody with a training in mathematical physics. The part $B_\phi \mathbf{e}_\phi$ is called the toroidal magnetic field, whereas the remaining part $B_r \mathbf{e}_r + B_\theta \mathbf{e}_\theta$, of which the field lines can be drawn in the poloidal plane, is called the poloidal magnetic field.

11 Babcock, H. W., & Babcock, H. D., 1955, 'The Sun's magnetic field, 1952–1954', *Astrophysical Journal*, **121**, 349–366.

Chapter 3: Here Comes the Sun

1 Schwarzschild, K., 1906, 'Über das Gleichgewicht der Sonnenatmosphäre', *Narchrichten von der Gesellschaft der Wissenschaften zu Göttinger*, 41–53.

2 Chandrasekhar, S., 1931, 'The maximum mass of ideal white dwarfs', *Astrophysical Journal*, **74**, 81–82.

3 The brief account of the Eddington–Chandrasekhar debate given here is based on Wali, K. C., 1991, *Chandra* (University of Chicago Press), pp. 124–127.

4 Davis, R., Harmer, D. S., & Hoffman, K. C., 1968, 'Search for neutrinos from Sun', *Physical Review Letters*, **20**, 1205–1209.

5 Leighton, R. B., Noyes, R. W., & Simon, G. W., 1962, 'Velocity fields in the solar atmosphere. I. Preliminary report', *Astrophysical Journal*, **135**, 474–499.

6 Deubner, F.-L., 1975, 'Observations of low wavenumber nonradial eigenmodes of the Sun', *Astronomy and Astrophysics*, **44**, 371–375.

Chapter 4: The Fourth State of Matter

1 An excellent account of the history of how electromagnetism developed can be found in Gamow, *The Biography of Physics*, Chapter V, which I have drawn upon.

2 Saha, M. N., 1920, 'Ionisation in the solar chromosphere', *Philosophical Magazine*, **40**, 472–488.

3 I may mention that, according to current ideas of cosmology, ordinary matter makes up only a small fraction of the contents of the universe. This ordinary matter is mostly in the plasma state.

4 Feynman, R. P., Leighton, R. B., & Sands, M., 1964, *Feynman Lectures on Physics, Volume II* (Addison-Wesley), p. I-1.

5 Bennett, W. H., 1934, 'Magnetically self-focussing streams', *Physical Review*, **45**, 890–897.

6 The origin of the name magnetic Reynolds number will be clear when we introduce the ordinary Reynolds number in fluid mechanics later in Section 6.1.

7 Alfvén, H., 1942, 'On the existence of electromagnetic-hydrodynamic waves', *Arkiv för matematik, astronomi och fysik*, **29 B**, No. 2.

Chapter 5: Floating Magnetic Buoys

1 Chandrasekhar, S., 1952, 'On the inhibition of convection by a magnetic field', *Philosophical Magazine*, **43**, 501–532.

2 Weiss, N. O., 1966, 'The expulsion of magnetic flux by eddies', *Proceedings of the Royal Society of London. Series A*, **293**, 310-328.

3 Parker, E. N., 1955, 'The formation of sunspots from the solar toroidal field', *Astrophysical Journal*, **121**, 491–507.

4 Choudhuri, A. R., & Gilman, P. A., 1987, 'The influence of the Coriolis force on flux tubes rising through the solar convection zone', *Astrophysical Journal*, **316**, 788–800.

5 Moreno-Insertis, F., 1986, 'Nonlinear time-evolution of kink-unstable magnetic flux tubes in the convective zone of the sun', *Astronomy and Astrophysics*, **166**, 291–305.

6 Choudhuri, A. R., 1989, 'The evolution of loop structures in flux rings within the solar convection zone', *Solar Physics*, **123**, 217–239.

7 D'Silvs, S., & Choudhuri, A. R., 1993, 'A theoretical model for tilts of bipolar magnetic regions', *Astronomy and Astrophysics*, **272**, 621–633.

Chapter 6: Dynamos in the Sky

1 Reynolds, O., 1883, 'An experimental investigation of the circumstances which determine whether the motion of water in parallel channels shall be direct or sinuous and of the law of resistance in parallel channels', *Philosophical Transactions of the Royal Society of London*, **174**, 935–982.

2 See Chapter 1, Note 8, for the full reference.

3 Yoshimura, H., 1975, 'Solar-cycle dynamo wave propagation', *Astrophysical Journal*, **201**, 740–748.

4 Cowling, T. G., 1957, *Magnetohydrodynamics* (Interscience Publishers), p. 89.

5 I am indebted to Professor Karl-Heinz Rädler for providing the photograph of Max Steenbeck and for information about his life.

6 Steenbeck, M., Krause, F., & Rädler, K.-H., 1966, 'Berechnung der mittleren Lorentz-Feldstärke für ein elektrisch leitendes Medium in turbulenter,

durch Coriolis-Kräfte beeinflusster Bewegung', *Zeitschrift Naturforschung Teil A*, **21**, 369–376.

7 Steenbeck, M., & Krause, F., 1969, 'Zur Dynamotheorie stellarer und planetarer Magnetfelder. I. Berechnung sonnenähnlicher Wechselfeldgeneratoren', *Astronomische Nachrichten*, **291**, 49–84.

8 Gailitis, A., with 10 co-authors, 2000, 'Detection of a flow induced magnetic field eigenmode in the Riga dynamo facility', *Physical Review Letters*, **84**, 4365–4368.

Chapter 7: The Conveyor Belt inside the Sun

1 Babcock, H. W., 1961, 'The topology of the Sun's magnetic field and the 22-year cycle', *Astrophysical Journal*, **133**, 572–587.

2 Leighton, R. B., 1964, 'Transport of magnetic fields on the Sun', *Astrophysical Journal*, **140**, 1547–1562.

3 Wang, Y.-M., Nash, A. G., & Sheeley, N. R., 1989, 'Magnetic flux transport on the sun', *Science*, **245**, 712–718.

4 Wang, Y.-M., Sheeley, N. R., & Nash, A. G., 1991, 'A new solar cycle model including meridional circulation', *Astrophysical Journal*, **383**, 431–442.

5 Choudhuri, A. R., Schüssler, M., & Dikpati, M., 1995, 'The solar dynamo with meridional circulation', *Astronomy and Astrophysics*, **303**, L29–L32.

6 Durney, B. R., 1995, 'On a Babcock–Leighton dynamo model with a deep-seated generating layer for the toroidal magnetic field', *Solar Physics*, **160**, 213–235.

7 Chatterjee, P., Nandy, D., & Choudhuri, A. R., 2004, 'Full-sphere simulations of a circulation-dominated solar dynamo: Exploring the parity issue', *Astronomy and Astrophysics*, **427**, 1019–1030.

8 Yeates, A. R., & Muñoz-Jaramillo, A., 2013, 'Kinematic active region formation in a three-dimensional solar dynamo model', *Monthly Notices of the Royal Astronomical Society*, **436**, 3366–3379.

9 Hazra, G., Karak, B. B., & Choudhuri, A. R., 2014, 'Is a deep one-cell meridional circulation essential for the flux transport solar dynamo?', *Astrophysical Journal*, **782**, 93 (12 pp).

10 Ramachandran, R., 2013, 'Light on sunspots', *Frontline*, January 11 issue, pp. 114–120.

Chapter 8: A Journey from the Sun to the Earth

1 Sabine, E., 1852, 'On periodical laws discoverable in the mean effects of the larger magnetic disturbances', *Philosophical Transactions of the Royal Society of London*, **142**, 103–124.

2 Birkeland, K., 1908, *The Norwegian Aurora Polaris Expedition 1902–1903* (Christiania Oslo).

3 Størmer, C. 1931, 'Ein Fundamentalproblem der Bewegung einer elektrisch geladenen Korpuskel im kosmischen Raume', *Zeitschrift für Astrophysik*, **3**, 31–52.

4 Hess, V. F., 1912, 'Über Beobachtungen der durchdringenden Strahlung bei sieben Freiballonfahrten', *Physikalische Zeitschrift*, **13**, 1084–1091.

5 Fermi, E., 1949, 'On the origin of the cosmic radiation', *Physical Review*, **75**, 1169–1174.

6 van de Hulst, H. C., 1953, 'The chromosphere and the corona', in *The Sun*, ed. G. P. Kuiper (University of Chicago Press), p. 207.

7 Edlén, B., 1943, 'Die Deutung der Emissionslinien im Spektrum der Sonnenkorona', *Zeitschrift für Astrophysik*, **22**, 30–64.

8 Parker, E. N., 1958, 'Dynamics of the interplanetary gas and magnetic fields', *Astrophysical Journal*, **128**, 664–675.

9 Parker, E. N., 1972, 'Topological dissipation and the small-scale fields in turbulent gases', *Astrophysical Journal*, **174**, 499–510.

10 Alfvén, H., 1942, 'Existence of electromagnetic-hydrodynamic waves', *Nature*, **150**, 405–406.

11 Chapman, S., & Ferraro, V. C. A., 1930, 'A new theory of magnetic storms', *Nature*, **126**, 129–130.

12 Cowling, T. G., 1985, 'Astronomer by accident', *Annual Reviews of Astronomy and Astrophysics*, **23**, p. 15.

Chapter 9: We Look Before and After

1 Svensmark & Calder, *The Chilling Stars*, p. 91.

2 Although systematic direct observations of the sun's polar magnetic field exist only from the mid-1970s onwards, there have been efforts to improve the statistics by inferring the value of the sun's polar magnetic field at earlier times from various indirect proxies. Although I have been involved in this research quite a bit, I refrain from getting into a discussion of this technical subject in this book.

3 Dikpati, M., & Gilman, P. A., 2006, 'Simulating and predicting solar cycles using a flux-transport dynamo', *Astrophysical Journal*, **649**, 498–514.

4 Tobias, S., Hughes, D., & Weiss, N., 2006, *Nature*, **442**, p. 26.

5 Choudhuri, A. R., Chatterjee, P., & Jiang, J., 2007, 'Predicting solar cycle 24 with a solar dynamo model', *Physical Review Letters*, **98**, 131103 (4 pp).

6 Jiang, J., Chatterjee, P., & Choudhuri, A. R., 2007, 'Solar activity forecast with a dynamo model', *Monthly Notices of the Royal Astronomical Society*, **381**, 1527–1542.

7 Dikpati, M., & Charbonneau, P., 1999, 'A Babcock–Leighton flux transport dynamo with solar-like differential rotation', *Astrophysical Journal*, **518**, 508–520.

8 Karak, B. B., & Choudhuri, A. R., 2011, 'The Waldmeier effect and the flux transport solar dynamo', *Monthly Notices of the Royal Astronomical Society*, **410**, 1503–1512.

9 Choudhuri, A. R., & Karak, B. B., 2012, 'Origin of grand minima in sunspot cycles', *Physical Review Letters*, **109**, 171103 (4 pp).

Chapter 10: Epilogue: Dynamos are Forever

1 Baliunas, S. L., with 26 co-authors (including Wilson, O. C.), 1995, 'Chromospheric variations in main-sequence stars', *Astrophysical Journal*, **438**, 269–287.

2 Schüssler, M., & Solanki, S. K., 1992, 'Why rapid rotators have polar spots', *Astronomy and Astrophysics*, **264**, L13–L16.

3 Glatzmaier, G. A., & Roberts, P. H., 1995, 'A three-dimensional self-consistent computer simulation of a geomagnetic field reversal', *Nature*, **377**, 203–209.

Suggestions for Further Reading

Non-technical popular science books

Blundell, S. J., 2012, *Magnetism: A Very Short Introduction* (Oxford University Press)

Brody, J., 2002, *The Enigma of Sunspots* (Floris Books)

Clark, S., 2007, *The Sun Kings* (Princeton University Press)

Clerke, A. M., 1885, *A Popular History of Astronomy during the Nineteenth Century* (Adam & Charles Black. Reprinted by Cambridge University Press)

Einstein, A., & Infeld, L., 1938, *The Evolution of Physics* (Simon and Schuster)

Gamow, G., 1941, *The Birth and Death of the Sun* (Viking Press. Reprinted by Dover)

Gamow, G., 1961, *The Biography of Physics* (Harper & Row. Reprinted by Dover as *The Great Physicists from Galileo to Einstein*)

Golub, L., & Pasachoff, J. M., 2001, *Nearest Star* (Harvard University Press)

Hoyle, F., 1955, *Frontiers of Astronomy* (Harper Collins)

King, A., 2012, *Stars: A Very Short Introduction* (Oxford University Press)

Lang, K. R., 2006, *Sun, Earth and Sky*, 2nd edn. (Springer)

Noyes, R. W., 1982, *The Sun, Our Star* (Harvard University Press)

Svensmark, H., & Calder, N., 2007, *The Chilling Stars* (Icon Books)

Uberoi, C., 2000, *Earth's Proximal Space* (Universities Press)

Zirker, J. B., 2009, *The Magnetic Universe* (Johns Hopkins University Press)

Advanced technical books

Alfvén, H., 1950, *Cosmical Electrodynamics* (Oxford University Press)

Chandrasekhar, S., 1961, *Hydrodynamic and Hydromagnetic Stability* (Oxford University Press. Reprinted by Dover)

Charbonneau, P., 2013, *Solar and Stellar Dynamos* (Springer)

Choudhuri, A. R., 1998, *The Physics of Fluids and Plasmas: An Introduction for Astrophysicists* (Cambridge University Press)

Choudhuri, A. R., 2010, *Astrophysics for Physicists* (Cambridge University Press)

Cowling, T. G., 1976, *Magnetohydrodynamics*, 2nd edn. (Adam Hilger)

Glatzmaier, G. A., 2013, *Introduction to Modeling Convection in Planets and Stars* (Princeton University Press)

Hanslmeier, A., 2007, *The Sun and Space Weather*, 2nd edn. (Springer)

Hoyt, D. V., & Schatten, K. H., 1997, *The Role of the Sun in Climate Change* (Oxford University Press)

Krause, F., & Rädler, K.-H., 1980, *Mean-Field Magnetohydrodynamics and Dynamo Theory* (Pergamon)

Lang, K. R., 2008, *The Sun from Space*, 2nd edn. (Springer)

Mestel, L., 1999, *Stellar Magnetism* (Oxford University Press)

Moffatt, H. K., 1978, *Magnetic Field Generation in Electrically Conducting Fluids* (Cambridge University Press)

Parker, E. N., 1979, *Cosmical Magnetic Fields* (Oxford University Press)

Parker, E. N., 2007, *Conversations on Electric and Magnetic Fields in the Cosmos* (Princeton University Press)

Phillips, K. J. H., 1995, *Guide to the Sun* (Cambridge University Press)

Priest, E. R., 1982, *Solar Magnetohydrodynamics* (Reidel)

Rüdiger, G., Kitchatinov, L. L., & Hollerbach, R., 2013, *Magnetic Processes in Astrophysics* (Wiley–VCH)

Schrijver, C. J., & Zwaan, C., 2000, *Solar and Stellar Magnetic Activity* (Cambridge University Press)

Stenflo, J. O., 1994, *Solar Magnetic Fields* (Kluwer Academic Publishers)

Stix, M., 2004, *The Sun*, 2nd edn. (Springer)

Thomas, J. H., & Weiss, N. O., 2008, *Sunspots and Starspots* (Cambridge University Press)

Vaquero, J. M., & Vázquez, M., 2009, *The Sun Recorded Through History* (Springer)

Zeldovich, Y. B., Ruzmaikin, A. A., & Sokoloff, D. D., 1983, *Magnetic Fields in Astrophysics* (Gordon and Breach)

Zirin, H., 1988, *Astrophysics of the Sun* (Cambridge University Press)

Index